# Cannabis Alchemy

# Cannabis Alchemy

## The Art of Modern Hashmaking

### Methods for Preparation of Extremely Potent Cannabis Products

## by D. Gold

**20TH CENTURY ALCHEMIST**

**Cannabis Alchemy**
ISBN: 0-914171-40-2
ISBN: 978-0-914171-40-9
Copyright 1973, 1989 by David Hoye

Published by Ronin Publishing, Inc.
Post Office Box 22900
Oakland, CA 94609
www.roninpub.com

| | |
|---|---|
| Cover Design | Bonnie Smetts |
| Cover Photo | Harlan Ang |
| Typesetting | Aardvark Type |
| Editor | Nicholas Flamel |
| Copy Editor | Stefan Ekeland |
| Paste-Up | Sandy Drooker & Phil Gardner |
| Index | Syre Van Young |

Distributed to the trade by Perseus/Publishers Group West
Printed in the United States of America

*Dedicated to Dr. Adams*

# Contents

# Preface

The cultivation of marijuana and the refinement of its preparations has concerned alchemists and hedonists on this planet for centuries. *Cannabis sativa* and *Cannabis indica* are both powerful allies. The body of the plant itself serves as a link between the physical plane and a host of Spirits of exceptional wisdom and subtlety. When the plant is ingested, these qualities are manifested in the mind of the worshipper, unlocking the storehouse of Wisdom within and revealing the hidden springs of pleasure. Smoking or eating the leaves or flowers is usually sufficient to bring about the desired state, although it seems inherent in the nature of Man to search for more concentrated forms of the drug that are stronger, more pleasant to ingest, or more desirable in some other way. Thus in every culture the technology of that period is applied to the work of the transmutation. As technology has evolved, so have the outward trappings of the operation, even though the principles underlying the operation remain constant throughout time and cultural differences. In primitive situations the refinement is carried on manually, the flowers being separated from the less psychoactive seeds, stems and leaves. Resins are extracted by simply rubbing the plant with the hands and then scraping the resin from the hands with the fingers. Water extractions are accomplished by boiling the plant parts in water, letting the water evaporate in the sun, and then collecting the residue. In cultures advanced to the state of mechanical technology, certain devices are used to this end. This might involve sifting the dried resin through mesh cloth, or mechanically pressing the resin into slabs. In cultures where the ingestion of the plant is accepted and desirable, these techniques become the formulae of power, and hashmakers are revered as Priests of the Holy Sacrament. In other situations their work is misunderstood, and they are branded as criminals to be persecuted. Yet their work continues.

*— D. Gold*

# Introduction

The public attitude on the personal use of marijuana has changed significantly since the first edition of this book. It appears to be only a matter of time before the legal status of marijuana becomes like that of tobacco (in the past, tobacco use was widely prohibited and its use was grounds for excommunication from the Catholic Church). The use of THC has been allowed in the treatment of glaucoma and to alleviate the unpleasant side effects of chemotherapy in the treatment of cancer. The reports of undesirable effects of marijuana do appear in the scientific literature, but there is no consensus on the damaging effects of the moderate use of marijuana. In contrast, the scientific consensus on damaging effects of smoking tobacco or using alcohol in excess is overwhelming. The logical explanation of this situation is convoluted and is probably understood fully only by those with a half-century of experience in social engineering and government. When the situation is presented to an individual being initiated to the ways of society, it only damages the credibility of the whole system. The real motivation for a change in governmental attitudes no doubt comes from the simple fact that the use of marijuana has become a common practice for millions of Americans, includ-

ing the sons and daughters of the governing officials—
the criminal stigma and revolutionary symbolism of
marijuana use has faded.

Given the widespread everyday recreational use of
cannabis, it becomes obvious that the various methods
of consuming it are only starting to develop. If one
walks down the aisle of a wine shop and considers the
varied forms of ethanol which have evolved, it is clear
that cannabis has a long way to go. The research on
the physiological and psychological effects of THC
and its analogs is in its incipient state, and the potential for developing compounds of varied effects and
duration is enormous. The pharmaceutical companies
are active in this area of research, and it is in their
interest that the use be decriminalized but subject to
prescription. The problem (for the regulators) is that
cannabis is simple to grow anywhere and the methods
for extracting and refining THC are an elementary
chemical technique. This volume is one of the first
reports of how simple techniques are being used to
create some simple variants of the cannabis experience. The compounds of the future can only be
guessed at, but if the past enterprise and energy of
alchemists is any indication, it won't be long before
we see for ourselves.

Nicolas Flamel
Berkeley, California

# Cannabis Alchemy

# One

# Extraction and Purification of Marijuana and Hashish Oils

### Step 1: Preparing the marijuana or hashish

If marijuana is to be used as a starting material, the seeds are removed prior to extraction. The remaining material is them crumbled or broken and the stems cut short with scissors. The marijuana is dried thoroughly. An oven is preheated to 250°F and turned off. The marijuana is placed on a cookie sheet for fifteen-minute intervals until the loose leaf and flower parts may be easily crumbled to powder between the palms. This procedure prevents scorching the plant.

Hashish may be heated for several minutes in an oven or in a frying pan at low heat until it begins to smoke slightly. It is then easily crumbled in the hands, or, if a mortar and pestle is available, it may be ground to a fine powder. Powdered hashish exposed to the air for long periods will decrease in potency, so this grinding is done just prior to extraction.

### Step 2: Pulverizing the cannabis material

There are several reasons for reducing the material to the finest powder possible. Ruptured cell walls allow the oil to be extracted more readily and the volume of the starting material is reduced, thus lessening the size of the extraction apparatus needed, as well as the amount of solvent required.

Marijuana is put into a heavy-duty blender until it is one-third full. A slower speed allows the ground material to fall into the blades while constantly flowing up the sides from the bottom. If necessary, the material can be agitated with a wooden stick while the blender is not running. It is dangerous to stir while the blender is operating, as the stick can be shot from the

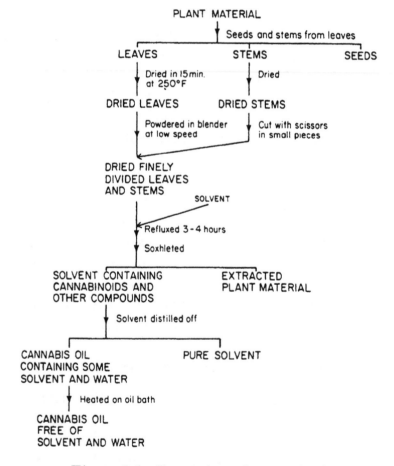

Figure 1.1  Extraction of cannabis oil.

blender with great force. It is easier to grind the chopped stems separately and then mix the powdered material before proceeding to the next step.

Hashish may also be ground in a blender, but small amounts should be run, as larger amounts will put a strain on the motor. Pressed forms of hashish may be shredded with a cheese grater prior to blending.

## Step 3: Refluxing

The essential oil is extracted from the cannabis material by refluxing (boiling) in a solvent. This essential oil (containing THC and related substances, chlorophyll, and the substances which contribute the taste and smell) dissolves in the solvent (usually an alcohol), while the cellulose parts of the herb do not dissolve. The leached marijuana is removed by straining, and the solvent containing the oil is evaporated, leaving as residue the essential oil of the herb.

As it is very dangerous to boil solvents (the fumes and liquid are quite flammable), it is necessary to use specialized methods in order to perform the operation safely.

Refluxing apparatus is composed of the following items:

1. A small pot, preferably of stainless steel, to hold the powdered cannabis material and solvent. The pot should not be over two-thirds full when the marijuana is covered with half again its volume of solvent.

2. A large stew pot with lid at least fifty per cent wider and twice as deep as the smaller pot mentioned above. Both pots should have flat bottoms.

3. A large, deep tub for boiling water, at least twice as wide as the stew pot.

4. A heavy-duty electric hotplate with two burners.

5. Several yards of one-inch hemp or manila rope.

6. Large, thick polyethylene trash bags. Three-mil-thick trash bags are best.

7. Innertube cut into one-inch-wide rubber bands to fit tightly around the stew pot.

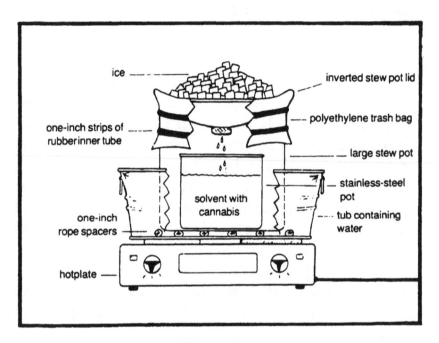

ice

inverted stew pot lid

polyethylene trash bag

one-inch strips of rubber inner tube

large stew pot

stainless-steel pot

solvent with cannabis

one-inch rope spacers

tub containing water

hotplate

Figure 1.2

**Assembly:** The large tub is placed securely on the hotplate. Lengths of one-inch rope are then placed in the bottom of the large tub to keep the stew pot (set in the tub on top of the rope) from resting directly on the bottom of the tub. The small stainless-steel pot containing the powdered cannabis and solvent is placed in the stew pot, and the lid is placed on top of the stew pot in an inverted position (upside down). A piece of plastic trash bag is cut and placed over the lid of the large stew pot so that it extends halfway down the side. The unit is sealed by securing the polyethylene sheet to the stew pot with two large innertube rubber bands. The innertube bands are positioned several inches down the side of the pot, allowing some slack in the plastic sheeting. Any air that is trapped under the plastic is forced out by loosening the rubber bands and flattening the bag with the hand. As much ice as possible is piled on the plastic bag covering the inverted lid of the stew pot. The tub is filled half full of water which is then brought to a boil. This heats the apparatus to about 212°F, but not above.

As the stainless-steel pot containing the cannabis and solvent is heated by boiling water in the tub, the solvent boils. As the fumes rise inside the apparatus, they make contact with the inverted lid of the stew pot, which is cooled by ice from above. These fumes of solvent then condense to liquid, relieving the pressure created by boiling, and drop off the inverted lid back into the stainless-steel pot containing the cannabis and solvent. Be refluxing in this manner, there is no danger of explosion or of toxic fumes escaping into the air.

The reason that the cannabis and solvent are not put directly into the large stew pot is that the condensing surface area (the ice-cooled lid) must be larger than the surface area of the boiling solution in the stainless-steel pot.

The plastic sheeting is used for several reasons. The reaction is completely sealed from the atmosphere, preventing any fumes from escaping or igniting. A rigid seal, such as the locking top of a pressure cooker, is not good, as it would prevent pressure build-up in the stew pot from causing the plastic bag to inflate. The inflation of the plastic notifies the chemist of the pressure increase and also causes the ice to fall into the boiling water bath, cooling the rig to a safe temperature and reducing pressure within the system. Pressure will not build up too high unless one neglects to keep enough ice on top or allows the apparatus to heat up too fast before the ice has sufficiently cooled the inverted lid. Refluxing is done for three or four hours. Most of the essential oils of the cannabis material are now dissolved in the solvent.

There are several solvents that work well. Their properties, and the advantages and disadvantages of each, are discussed below:

1. Methyl alcohol, methanol, wood alcohol (boiling point 64°C). This solvent is commonly employed and, if used correctly, does a fine job. Methanol is available at many pharmacies and in larger quantities at industrial chemical supply companies. It is also available as paint thinner, but it is seldom very pure in this form. Methanol fumes are toxic and explosive. Inhalation of these fumes makes one sick, with pronounced body ache. Continued inhalation of even small amounts may cause permanent damage. Any traces of the solvent remaining in the oil product will be hazardous to the consumer. Methanol evaporates at a uniform temperature (approximately 190°F) and does not extract a lot of the water-soluble tars, which are not psychoactive. A method for removing traces of the solvent will be discussed later.

2. Rubbing alcohol (most rubbing alcohol is 70% isopropyl alcohol and 30% water). There are several advantages to using isopropyl rubbing alcohol. It is

available in many stores at a low price and is much less toxic and explosive than methanol. Unfortunately, because it contains water, many of the water-soluble, non-psychoactive substances are also extracted. The oil yield using rubbing alcohol is twice that of methanol, and is proportionally less potent. Water-soluble tars may also give the oil undesirable taste and burning qualities. If the oil is to be re-extracted later with a more selective solvent, however, it matters little what it is like at this point. The water in the mixture also causes it to evaporate at a much higher temperature than methanol. Once the alcohol is completely evaporated, the water that was in the solvent remains with the oil. This takes a long time to evaporate in a boiling-water bath. An oil bath may be used. The temperature of the oil in the bath is kept slightly higher than the boiling point of water. Water that gets into the oil bath may spatter; this is a hazard.

3. Ethanol, ethyl alcohol, pure grain alcohol (boiling point 78.5°C). This is a very desirable solvent. It has extraction properties very similar to methanol, but is not as toxic. It is very difficult to obtain, however, as it is a major active ingredient in liquor and is heavily taxed. Pure ethanol may be produced from either liquor or fermented material. Denatured ethanol, which is available in hardware stores and pharmacies, contains non-removable poisons which evaporate at the same temperature as pure ethanol. This makes the ethanol unfit for drinking.

4. Petroleum ether (boiling point 30–60°C). Petroleum ether is a light solvent much more selective than any of the alcohols. Extracting with petroleum ether produces an oil that is twice as potent by weight as oil

extracted with alcohol. The cannabis material may be extracted directly with ether but, due to petroleum ether's highly explosive nature, the oil is first removed from the plant material with alcohol and then re-extracted with ether. This requires a much smaller amount of the dangerous solvent. Petroleum ether is usually available only through chemical supply companies.

**Step 4: Soxhleting**

After refluxing, it is necessary to remove the oil-bearing solvent that still remains in the expended cannabis material. This is done by draining the dark oil/solvent liquid from the cannabis material and wash-

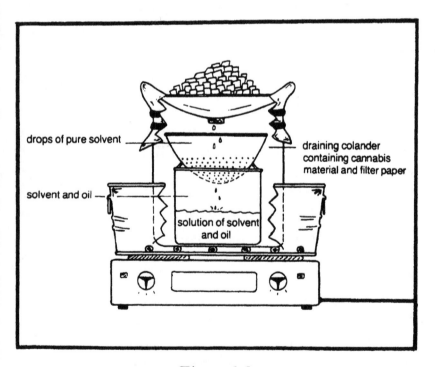

drops of pure solvent

draining colander containing cannabis material and filter paper

solvent and oil

solution of solvent and oil

Figure 1.3

ing the material repeatedly with clean solvent. A vegetable-draining colander is placed above the stainless-steel pot mentioned before. The colander is fitted with a large coffee filter paper (twelve-inch David Douglas brand papers are quite adequate) and the cannabis/solvent/oil mixture is poured into the colander. The oil-bearing, dark-colored solvent/oil mixture drains from the bottom of the colander, free from particles of vegetable matter. The colander is then set on top of the stainless-steel pot, now containing the alcohol/cannabis-oil solution, and the apparatus is reassembled in the same manner as for refluxing.

As the solvent/cannabis-oil solution boils, the alcohol fumes rise until they meet the ice-cooled lid and recondense into liquid. The oil does not evaporate and remains in the stainless-steel pot. The drops of pure, recondensed solvent fall from the ice-cooled lid and drip through the colander containing the cannabis material. The oil remaining in the cannabis material is washed out and drains into the stainless-steel pot. The oil is totally extracted when several drops of the liquid draining from the colander leave no colored residue when evaporated on a piece of glass. Before opening the apparatus after soxhleting, the rig is cooled sufficiently to condense any fumes. Setting the stew pot in a tub of ice and water for several minutes is one method of doing this. A thick blanket can be kept soaking in the tub. This is an excellent safety measure, since a water-soaked blanket is an excellent fire extinguisher.

**Step 5: Removal of the solvent from the oil**

To distill off the solvent, a small collection pan re-

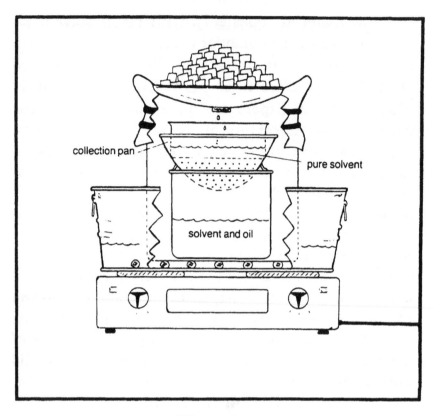

collection pan

pure solvent

solvent and oil

Figure 1.4

places the expended cannabis material in the colander (which may be discarded). The apparatus is then reassembled and returned to the water bath. The solvent/ oil solution in the small stainless-steel pot boils; the fumes rise and are condensed on the ice-cooled lid as before. The pure solvent drips into the colander, where it is collected in the small pan. The oil remains behind in the stainless- steel pot. The collected solvent, which is essentially pure, may be saved for a future extraction.

After the solvent is removed and collected, the stainless-steel pot containing the oil is kept in boiling water to remove all traces of the solvent. If a toxic solvent, or one containing water is used, steps are taken to remove the last traces of solvent and water. Some water is added to the oil and evaporated in an oil bath (cottonseed oil works fine) at approximately 220°F. When the water is gone, all traces of the solvent have been removed, since all solvents mentioned here evaporate at a temperature below that of boiling water. The oil may now be eaten or smoked.

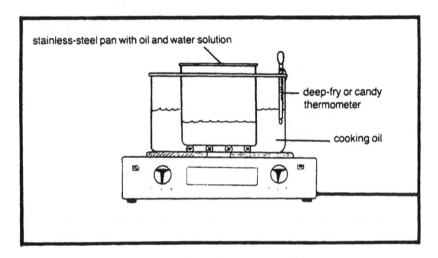

Figure 1.5

## Step 6: Purification

Oil produced using this method is quite potent, but still contains substances which give the oil its taste, smell, and color. These are sometimes very pleasant to smoke, and they are sometimes left in the oil. Removing them, however, greatly increases the potency, but decreases the yield proportionally.

The oil from the extraction is dissolved in five times its weight of alcohol and poured into an equal volume

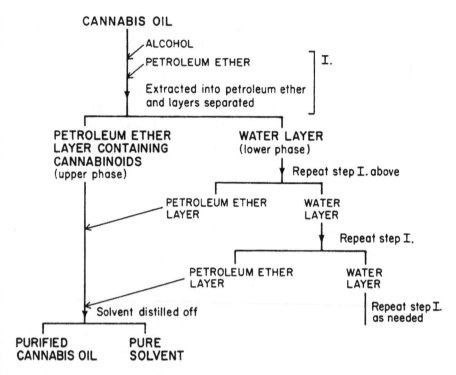

Figure 1.6 Purification of cannabis oil.

of water in a large glass jug with a screw-on cap. All solvents used are cold or cool. A volume of petroleum ether equal to half the volume of the water used is added. The screw top is tightened and the jug is inverted. The jug is turned upright immediately, and a few seconds later, when the mixture has run down the sides of the jug, the screw cap is opened slowly to relieve the pressure. The inversion of the jug is repeated about twenty-five times, which releases the pressure each time, and then the jug is allowed to sit for about half an hour. The mixture of liquids will separate into three distinct layers. The bottom layer will contain water, alcohol, and the substances in the oil (tars and resins) that are not soluble in petroleum ether. The thin middle layer is an emulsion of waxes, ether, and air bubbles. The top layer is the purified oil dissolved in petroleum ether.

The jug is fitted with a two-holed rubber stopper, glass tubing, and rubber hose. Two pieces of glass tubing are fitted into the two-holed rubber stopper. (Injuries when cutting and fitting glass tubing are frequent—cut ends are always fire-polished and hands must be protected when any force is used in fitting the tubing.) One piece of tubing need only protrude from the stopper an inch on each side. The other tube is positioned so that when the stopper is

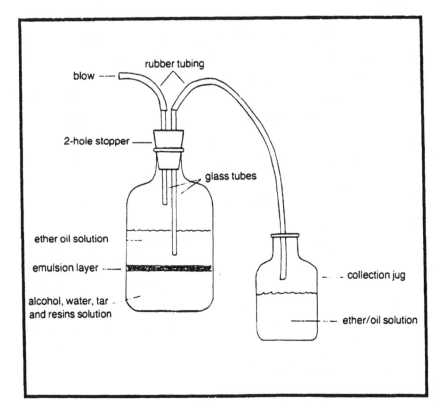

Figure 1.7

tightly fitted to the jug, the tube extends into the mixture to a half-inch from the bottom of the ether/oil layer. The other end of the tube is attached to a length of rubber tubing to transfer the ether/oil solution to a collection jug. The end of the tubing must be lower than the end of the tube in the ether/oil solution in order to obtain a siphon action. A short piece of rubber tubing is fitted to the short glass tube, and air pressure is applied to start the siphon.

The glass tube is spaced a bit above the emulsion layer; any ether/oil solution not removed will be recovered later. The ether/oil solution in the collection jug is saved. Another volume of fresh petroleum ether is added to the extraction by inverting, separating, and collecting the ether solution, which is added to the first ether/oil solution. This process is repeated until the ether layer remains clear after inverting. This indicates that the ether-soluble oil is totally extracted from the alcohol-and-water layer. No more than several ounces of the combined ether/oil solution is put in the stainless-steel pot, and the collection pan is placed in the colander. The apparatus is reassembled, as was done for the removal of the solvent from the oil after soxhleting. The rig is placed in the water bath and slowly heated to 140°F. After evaporating and collecting the ether (saved for future use), the stainless-steel pot with the oil is put into a boiling-water bath for several minutes and stirred occasionally to remove any residual traces of solvent. The refined oil thus obtained is much superior to the oil obtained from the original alcohol extraction.

# Two
# Isomerization

The oil produced by alcohol extraction and purification with petroleum ether contains tetrahydrocannabinol, two other compounds closely related to THC but non-psychoactive (cannabidiol and cannabinol), and several other compounds which contribute the taste and smell of the oil. The quality and quantity of the THC in the oil is determined by the quality and potency of the starting material. The oil from very strong cannabis material contains a much higher percentage of THC than the oil from marijuana or hashish that is less potent. The quality of the THC and the characteristics of the effect (high) are determined by the relative

(*Text continues on page 20.*)

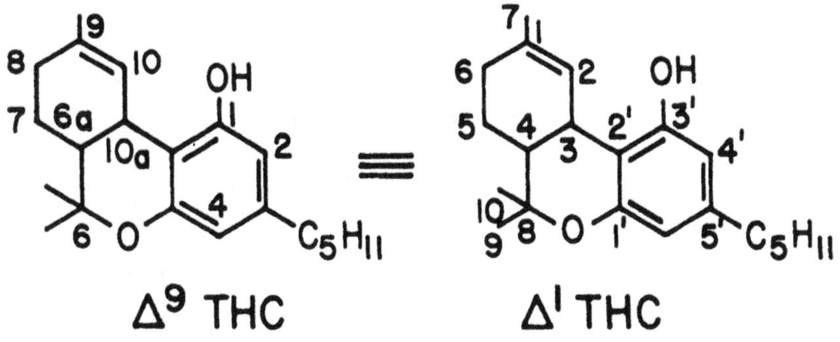

$$\Delta^9 \text{ THC} \equiv \Delta^1 \text{ THC}$$

Figure 2.1A  $\Delta^9$ THC is formal numbering and $\Delta^1$ THC is monoterpenoid numbering.

17

(1)

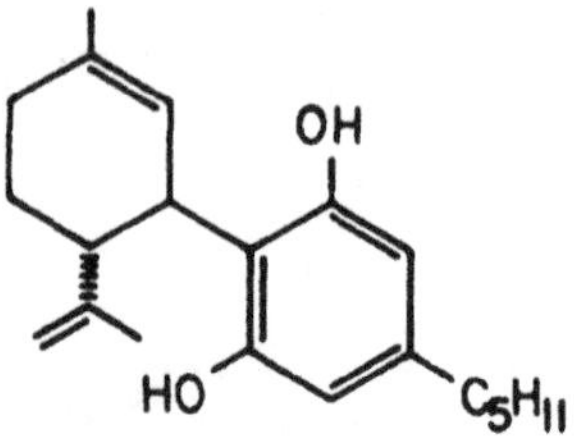

Cannabidiol

$H^+$ | Heat

(2)

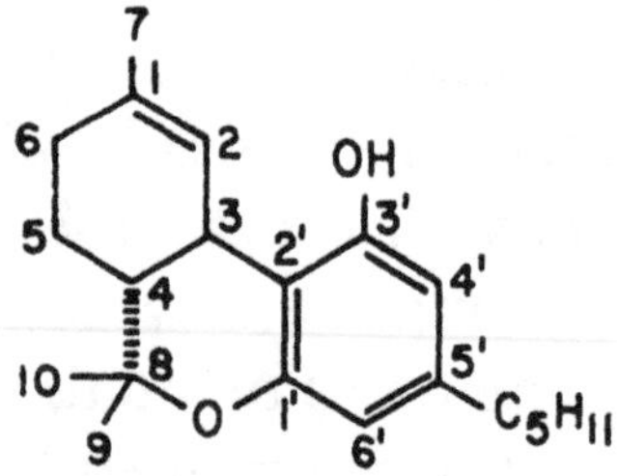

$\Delta^1$ – 3,4 – trans – tetrahydrocannabinol

$[\alpha]^{CHCl_3}_D - 150°$

$H^+$ | Heat

(3)

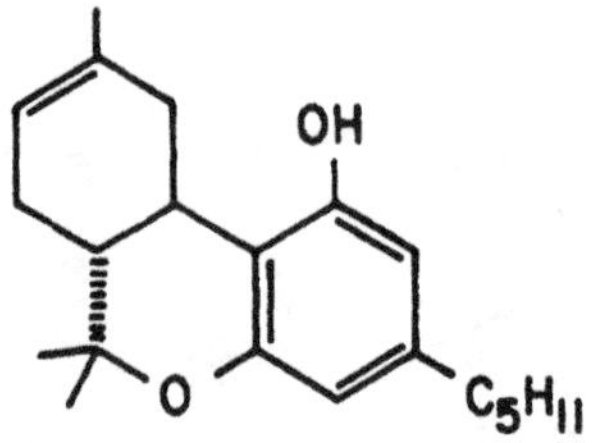

$\Delta^6$ – 3,4 – trans – tetrahydrocannabinol

$[\alpha]^{EtOH}_D - 260°$

Figure 2.1B  Many readers, especially those with some exposure to organic chemistry, may be curious about what

(*Caption continued from page 18.*)

happens to the molecules during the isomerization. The structural formulas for the reaction are given above (Fig. 2.1B). For the non-organic chemist, this is an opportunity to overcome your fear of formulas. The first credits for finding this reaction go to Roger Adams, who first described it in 1940; a detailed account appeared in 1941 (Adams *et al.*, 1941), but the correct formulas were not discovered until the 1960s. The changes that go on are minor; the first reaction forms a third ring by attaching the oxygen at 1' to the carbon at 8. The second reaction is just a shifting of one double bond from the 1,2 position to the 6,1 position. Two conventions for numbering the ring system are used in the literature; sometimes $\Delta^1$ is called $\Delta^9$ and $\Delta^6$ is called $\Delta^8$ (Mechoulam, 1970). See Figure 2.1A.

Since two different THCs are products of the reaction, the question becomes: how much of each is produced? This depends on the reaction conditions. If more concentrated or stronger acid is used and the time of reaction is increased, more $\Delta^6$ THC is produced; there is a 90% conversion of $\Delta^1 \rightarrow \Delta^6$ THC if *p*-toluenesulfonic acid in toluene is used for 10 hours at 100°C (Hively *et al.*, 1966). With mild conditions (absolute ethanol, 0.05% hydrochloric acid boiled for 2 hours) the product is mainly $\Delta^1$ THC (Gaoni and Mechoulam, 1964). So, mild conditions give $\Delta^1$ THC; more vigorous conditions give $\Delta^6$ THC.

This sounds simple enough; in reality, however, other factors can play a significant role. Starks (1977) mentions some side reactions, which have been reported; these vary with the solvent and conditions used for the isomerization. There is recent evidence (Bonuccelli, 1979) that solvents such as chloroform which promote free-radical reactions can result in large decreases in THC content. A chloroform solution of marijuana exposed to sunlight for 30 minutes lost 25–35% of its THC; the reaction was slower in the dark but still significant. The authors recommend ethanol over chloroform, and cool, dark storage. Slight decomposition even occurs with ethanol.

Given that $\Delta^1$ and $\Delta^6$ THC are both possible products, another important question is whether there is any dif-

*(Caption continued from page 19.)*

ference in the pharmacological effects between the two. Mechoulam (1970) says workers have found them to be about equal in activity, but some differences have been reported. Most studies are carried out on laboratory animals, so the effects measured are not mainly psychological. Psychological effects can be subtle, and the possibility of a difference between the effects of $\Delta^1$ and $\Delta^6$ THC are yet to be researched. The methodology for measuring such subtle changes remains a challenge to research psychologists.

positions of the double bonding in the THC molecule. The higher-rotating forms are more potent than the low-rotating and produce a higher, more psychedelic and spiritual effect. Methods for converting THC from low- to high-rotating follow.

The quantity of cannabidiol in the oil is important, as it may be converted to THC, thereby increasing the potency of the oil proportionally. Experience has indicated that the quantity of cannabidiol is usually at least equal to the quantity of THC. Because of this, the strength of the oil can be at least doubled through isomerization, and in some cases potency may be increased five to six times.

By using the correct chemicals and methods to convert the cannabidiol to THC, it is possible simultaneously to convert the THC (that which occurs naturally in the oil and also that which has been produced from cannabidiol) to higher-rotating forms. The highest benefit is obtained by starting with material high in cannabidiol, isomerizing the cannabidiol to THC, and converting the THC to its higher-rotating form. Both the potency of the oil and the quality of the high are greatly increased. The operation is carried out as follows:

The oil from the ether extraction is dissolved in absolute ethanol or pure methanol in the ratio of one gram of oil to ten grams of solvent. The ethanol may be denatured, but must not contain water. One drop of 100 per cent sulfuric acid is added to the alcohol/oil solution for each gram of oil. The acid is added slowly with continuous stirring. Pure sulfuric acid is very strong and will cause severe burns. Safety glasses, long rubber gloves, and clothing that covers as much of the body surface as possible are advised when working with it. Sulfuric acid burns are treated by immediate washing with water and bicarbonate of soda. The sulfuric acid is kept in a safety bottle made by permanently fitting a glass bottle with a screw top in a styrofoam-lined metal can.

A Pyrex pot containing the oil/alcohol/sulfuric acid solution is placed in the refluxing apparatus originally used for refluxing the material in alcohol. Pyrex is substituted for the stainless-steel pot because of the reactive nature of the sulfuric acid. The rig is placed in the boiling water bath and refluxed for two hours. At the end of this time, the stew pot is placed in an ice-water bath and opened. The solution is poured into an equal amount of water and extracted with petroleum ether, as was done in removing the ether-soluble oil from the alcohol extract solution. The ether solution is then poured into four volumes of water and gently inverted twenty-five times, releasing the pressure each time. The layers are allowed to separate, the ether/oil layer is siphoned off and the water is discarded. The ether/oil solution is poured into four volumes of 5% bicarbonate of soda solution in water. It is mixed, then separated, and the ether/oil

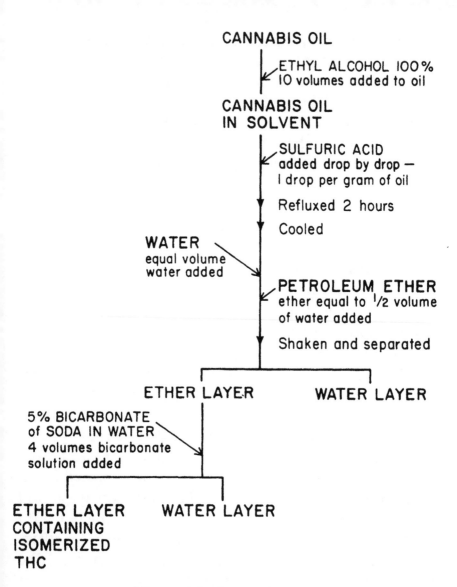

Figure 2.2  Isomerization.

layer is siphoned off. The bicarbonate of soda solution is discarded and the previous step (washing with pure water) is repeated twice. The ether is evaporated from the ether/oil solution, as was done previously in the first purification, using the stew pot apparatus. The pure ether is collected in the pan held in the colander. The oil now contains a much higher percentage of THC (determined by the amount of cannabidiol originally present). The THC is of the high-rotating isomeric form, and all of the toxins have been removed from the oil.

# Three

# THC Acetate

THC acetate has twice the potency of THC. On the Adams scale THC = 7.3, while its acetate = 14.6. Furthermore, there is a 25% increase in weight after adding the acetate structure. The effect of the acetate is more spiritual and psychedelic than that of the ordinary product. The most unique property of this material is that there is a delay of about thirty minutes before its effects are felt.

### Building a safety box in which to convert high-rotating THC to its acetate

Because this conversion utilizes a very dangerous chemical, acetic anhydride, a safety box is constructed in which to perform this operation. This places a shield between the chemist and the apparatus, and the operation takes place in a separate atmosphere. The fumes from heated acetic anhydride are very flammable and poisonous. Inhalation of the fumes is a most unpleasant and dangerous experience, so a glove box protects the operator from any contact with the fumes. Acetic anhydride is so difficult to handle safely that it is a necessity to use standard laboratory equipment and procedure. The reaction is monitored and controlled from outside the box by observing the equipment through a safety-glass window and manipulating the apparatus with long gloves sealed to the shield. The box is equipped with an adequate exhaust fan with a sparkless electric motor to quickly evacuate any fumes that arise while transferring solutions or from a spill or other mishap. A fire extinguisher is mounted inside the box. In case of fire or explosion, the chemist is protected by the thick front piece of the box and by its structural design. The box can also be used for any other chemical operation requiring an artificial atmosphere to avert fire or explosion.

An artificial atmosphere is created by replacing the air in the sealed box with anhydrous nitrogen gas. This makes flame or combustion (oxidation in general) impossible. The nitrogen is introduced into the chamber through an opening in one side, near the top of the box. The displaced air is removed through a valve near the bottom of the opposite end. Success here is determined by attempting to strike a match inside the box. When oxygen has been removed, this becomes impossible. The exhaust fan is then used only when it becomes necessary to evacuate the atmosphere in the box.

The three stations are utilized as follows: The equipment and bottled chemicals are put into the right-hand side of the box by lifting the hinged side piece. The apparatus is prepared at this station and operated in the middle at station two. The bottles are opened at station one, using the gloves, from the outside of the box after the side flap is closed and the atmosphere has been replaced with anhydrous nitrogen gas. The operator need never be exposed to the dangerous chemicals except within the controlled atmosphere of the box.

Diagrams showing details of construction are shown below:

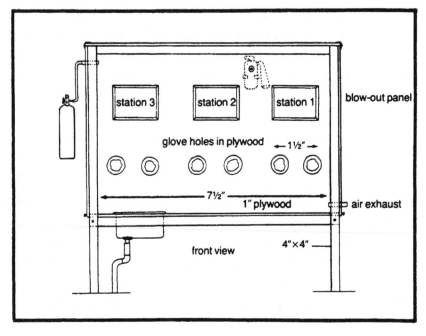

Figure 3.1

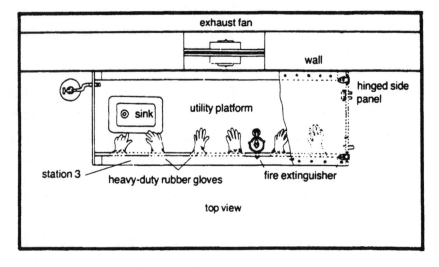

Figure 3.2

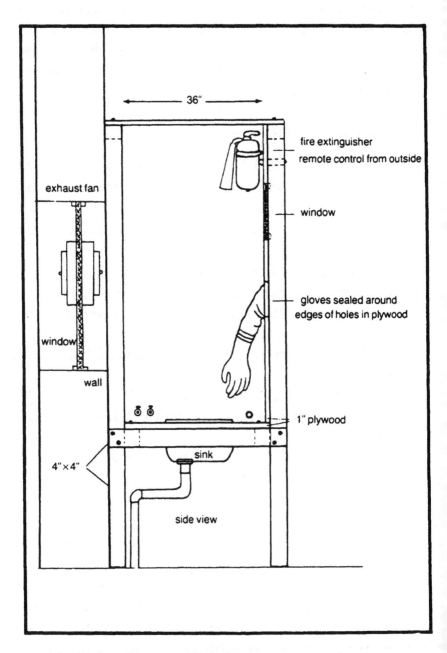

Figure 3.3

The side walls of the box are made of thin plywood and hinged at the top. This serves two purposes: access to the box is available from both ends, and in case of explosion, the force would be expended through the side panels, while the thick, reinforced front boards protect the chemist.

## Conversion of THC to its acetate

THC is converted to THC acetate by refluxing for two hours with acetic anhydride. The following apparatus is assembled as illustrated in the diagram:

1. 500 ml Pyrex round-bottom boiling flask with a ground glass fitting.

2. Tubular type condenser with ground glass male fitting that matches the fitting on the boiling flask.

3. Metal pot of at least 2000 ml as hot oil bath for heating boiling flask.

4. Thermometer for monitoring the oil bath temperature.

5. Sparkless electric hotplate.

6. Rheostat to control temperature of hotplate from outside the box.

7. Ring stand and proper clamps for securing flask and condenser.

8. Cottonseed oil.

9. Acetic anhydride.

10. Immersible water pump, bucket, and hoses for filling condenser.

The principle of the refluxing operation is approximately the same as was used for isomerizing the cannabidiol to THC with the kitchen apparatus. The explosive and noxious nature of the acetic anhydride necessitates the use of the safety box. Although a

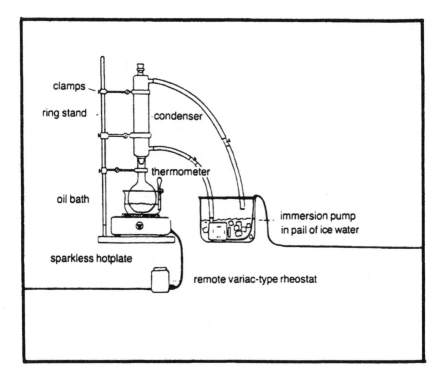

Figure 3.4

glove box is unnecessary for the operations of extraction and isomerization using the kitchen method previously described, these steps may also be done in the box as an added measure of safety.

The solution of acetic anhydride and cannabis oil is boiled in the round-bottom flask. The fumes rise into the icewater-cooled condenser, where they are condensed back into liquid, thus relieving the pressure created by boiling. The drops then fall back into the solution.

Before assembling the apparatus, these factors are taken into account: the temperature of the hotplate must be controlled from the outside of the box. This necessitates a variac-type rheostat in the power line to the hotplate. The pail containing the immersion pump which circulates the icewater coolant through the condenser should also be outside the box. There are two small holes in the safety box for the icewater input and return hoses. Although the sink at the left station of the box seems handy for the coolant pump, this would necessitate opening the side panel while the refluxing is in progress to add ice and remove water.

The equipment is assembled and operated in this manner: the right side panel is opened, and at station one the boiling flask, condenser, oil bath, and hotplate are assembled as illustrated in Figures 3.1 and 3.2. Each item is secured to the ring stand with adequate clamps. The flask is positioned at least one-half inch above the bottom of the oil bath. Electrical connections are not made nor the coolant hoses attached to the condenser yet, as the entire apparatus will be moved over to the center station before beginning.

The boiling flask, prior to being put in the safety box, contains a measured amount of cannabis extract. In the safety box are also placed the following: an unopened bottle of acetic anhydride, an empty graduated beaker, a beaker containing sufficient cottonseed oil to fill the oil bath to a level slightly above that of the cannabis oil/acetic anhydride solution that will be in the flask, and an empty, open-top container of the same height as the boiling flask and of a slightly larger diameter. This container holds the boiling flask safely when the apparatus is dismantled. The clamps holding the condenser are loosened and slid up the ring stand so that the mouth of the boiling flask is accessible for addition of the acetic anhydride. The right side panel is closed and the chamber is filled with nitrogen; then, using the gloves through the front board at station one, the bottle of acetic anhydride is opened. An amount is poured into the graduated beaker equal to three times the volume of the cannabis oil in the boiling flask. The cap on the acetic anhydride bottle is replaced and the acetic anhydride is poured carefully from the graduated beaker into the boiling flask. The condenser is securely replaced on the boiling flask and a solid rubber stopper is loosely fit to the top of the condenser.

The side panel is opened and the apparatus is moved to the center station. The input icewater hose is connected to the lower fitting of the condenser. The return hose runs from the uppermost fitting (assuring that the condenser is always filled with circulating water) through a hole in the safety box to a pail containing the icewater and immersion pump. The fittings are secured onto the condenser with twisted wire or automobile-type hose clamps. The power wire for the hotplate is also run through a hole in the safety box and connected to the rheostat. The oil bath temperature is monitored with a thermomoter, which is adjusted for observation through the safety glass window. The cottonseed oil is added to the bath. The empty beaker and the closed bottle of acetic anhydride are removed. The side panel is closed and secured. More nitrogen is now bled into the box until the air has been completely replaced by nitrogen. Striking a match is not a good idea for a test at this time, but if the test is tried a few times prior to beginning the operation, the time it takes to drive out the air completely can be estimated.

The pail containing the immersion pump is filled with water and ice, the pump is turned on, and the condenser is filled with circulating icewater. The oil

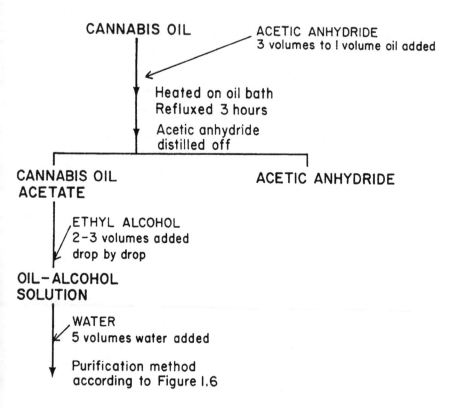

Figure 3.5 THC acetate synthesis.

in the oil bath is heated by turning on the electric hotplate. The temperature is raised slowly (indicated on the thermometer), giving the solution in the flask time to heat the temperature of the bath. The temperature is noted at which the solution of cannabis oil and acetic anhydride begins to fume and droplets of pure acetic anhydride form in the condenser and fall back into the solution. The temperature is slightly raised until the solution in the flask begins to boil. The bath temperature is stabilized at this point. This is continued for three hours. Ice is supplied as needed to the container with the immersion pump.

After three hours of refluxing, the electricity to the hotplate is turned off and the solution is allowed to cool to room temperature. Ice water is kept circulating through the condenser. After the solution has remained at room temperature for at least two hours, the rubber stopper at the top of the condenser is checked. It should form a perfect seal, but not be too tightly jammed into the condenser. The immersion pump is then turned off and the apparatus is allowed to sit at room temperature for another hour. At the end of this time the clamps holding the condenser are loosened and slid at the ring stand as before, giving access to the aperture of the boiling flask. The rubber stopper is removed from the top of the condenser and fitted tightly in the top of the boiling flask. The clamp holding the boiling flask is loosened and the flask is removed from the oil bath. The flask is wiped clean of oil and set into the empty open-top container set in the box earlier. The side panels are opened and the equipment is dismantled.

Removing the acetic anhydride by distillation is the next step (see Figure 3.6). The distillation requires the following equipment not used for refluxing:

1. A Pyrex distillation flask of the same capacity as the boiling flask, and two Erlenmeyer flasks, also of the same capacity.

2. An assortment of glass tubing, flexible tubing, and rubber stoppers.

3. A large pan to be used as an icewater bath for the Erlenmeyer flasks.

4. Several more ring stands and equipment clamps.

As the solution in the distillation flask is heated, the acetic anhydride vaporizes; the fumes rise and travel down the side arm of the distillation flask into the condenser, where they are cooled to liquid. The re-condensed acetic anhydride is collected in the receiving flask at the end of the condenser. This flask and a back-up flask used for safety are immersed in an ice-water bath.

The equipment is assembled at stations one and two as illustrated in Figure 3.6. The condenser is at a great enough angle that no acetic anhydride can lie between the condenser bottom and the exit tube. The glass tube for the introduction of the recondensed acetic anhydride extends deeper into the flask than the exit tube. The same is true of the back-up flask, even though it is unlikely that any fumes or liquid will go

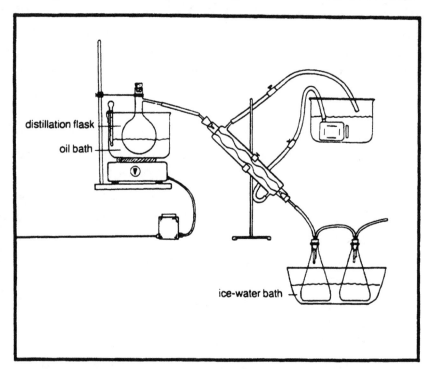

Figure 3.6

beyond the first flask. The tube leading from the back-up flask is open at the end.

The side panel is closed, the atmosphere is replaced with nitrogen gas as before, and the following process is used: using the gloves, the cannabis oil/acetic anhydride solution is poured from the boiling flask into the distillation flask. A funnel with a tube long enough to extend past the side arm of the distillation flask is used. This eliminates the chance of any solution running into the condenser. The rubber stopper is secured to seal the top of the distillation flask. The cottonseed oil is added to the oil bath, and the ice and water to the ice bath. The immersion pump and the hotplate are turned on. The temperature indicated by the thermometer in the oil bath is raised slowly to that used for the refluxing. This temperature is maintained until no more acetic anhydride is evaporated and collected. The volume of cannabis oil acetate now in the distillation flask will be up to twenty-five per cent more than the volume of the oil prior to acetylation. The oil temperature is maintained for one hour after the last traces of acetic anhydride have been removed. The hotplate is turned off, and, with the water still circulating through the condenser, the oil is allowed to cool to room temperature. The water from the ice bath is drained and the basin is wiped completely dry. Using the gloves at station one, the two-holed stopper is removed from the Erlenmeyer flask and replaced with a solid rubber stopper. The clamp is loosened and the flask is removed from the ice bath. The outside of the flask is thoroughly dried, and any traces of water are removed. The acetic anhydride is poured carefully from the flask into a safety container like that used for the sulfuric acid, a glass bottle fitted into a metal can.

The flask containing the cannabis oil acetate is removed from the safety box. Slowly, one drop at a time, several volumes of pure alcohol are added to dissolve the oil. This solution is poured into five volumes of water and extracted with petroleum ether, as was done in the purification techniques following the isomerization. The ether is evaporated in the stew pot appartus as before and collected. The resultant oil is redissolved in alcohol and poured once again into water. It is again extracted with petroleum ether, which is evaporated and collected as before. The resultant oil contains THC acetate and may be consumed in any of the customary manners.

# Four
# **Preparation of Hashish**

Hashish may be prepared from the extracted cannabis oil by mixing it with finely-powdered marijuana. The oil may be at any stage of refinement. Extremely strong hashish is obtained by using oil which has been isomerized, acetylated, and refined through removal of non-psychoactive compounds. Along with the potency of the oil itself, the ratio of oil to powdered marijuana determines the strength. In order for the hashish to be the proper consistency, a minimum of fifteen per cent oil must be used. This gives a product with the same consistency as powdery Moroccan or Lebanese hash. Fifty to sixty per cent oil (about equal parts of oil and powder) is the maximum amount of oil that can be used to give a product with hashish consistency. This product will be very strong and resemble in appearance and consistency the sticky, pliable *charas* of Nepal and India.

Powdered marijuana of the finest consistency is obtained by the following method: clean, very dry marijuana is pulverized in a high-speed blender. The material may be dried in a preheated oven (250°F) for fifteen-minute intervals. A very fine dust will collect on the blender top. This is sifted through a piece of nylon stocking or a very fine mesh screen.

Many times the taste of the hashish is improved if the oils giving the marijuana its taste and smell are removed from the dust. This is accomplished by extracting it with alcohol in the stew-pot apparatus as described earlier. Further extraction of the compounds contributing to the taste and smell is accomplished by boiling in water. All solvent is removed from the dust first, as the fumes might present a fire hazard. The water is filtered from the dust and the process is repeated with fresh water until the water remains clear, indicating that all soluble substances have been leached from the cellulose material. The cannabis dust is thoroughly dried, and is then ready to be mixed with the oil.

The mixing is facilitated by first heating the dust and oil and then working them together in a large mortar, or by kneading the mass with the hands. Thin, flat, hand-pressed patties like those from Afghanistan may be fashioned, or one may mold clumps of "fingers" or round "temple balls" such as those found in Nepal. Flat sheets and blocks may be formed by pressing the mixture between two heated steel plates in a vise.

# Five

# Increasing Potency of Intact Marijuana Flowers

The cannabis material is refluxed in the same manner as was done with the finely-powdered cannabis material previously, except that when processing intact flowers the material is first placed in a cheesecloth bag. The oil is then extracted from the marijuana in the usual manner. The oil is purified by re-extraction with petroleum ether and then isomerized and acetylated. The tars and resins left behind from the ether extraction remain dissolved in the alcohol/water layer in the extraction jug. The alcohol is evaporated and collected in the usual manner, and the water is evaporated in an oil bath at 220°F. The tars and resins thus obtained are mixed with the intensified, purified oil and dissolved in the exact amount of alcohol that the completely dry flowers will absorb. This amount is determined by adding clean alcohol to the dried flowers until they will absorb no more alcohol, but there is none lying in the bottom of the pan. The saturated flowers are then put into a distillation apparatus and all the solvent is removed and collected. This amount of alcohol is then mixed with the purified, intensified oil and the tars and resins. Using an oven-baster-type syringe, the flowers are equally saturated with the oil-bearing solvent. The saturated flowers are then put into the appropriate apparatus and the solvent removed. A small amount of water is then sprayed on the flowers. A steam iron or Sears wrinkle remover works fine. They are then placed in an oven which has been preheated to 250° and then turned off. Since the solvent evaporates at a much lower temperature than the water, when the flowers begin to dry out no traces of solvent will remain. The flowers are now coated with the intensified oil and may be over twelve times their original potency.

## Six

# Preparation of Oil Capsules

Capsules of oil for oral ingestion (sometimes called pot pills) are prepared by first mixing the purified oil with an equal amount of butter. The butterfat carries the oil through the membranes of the stomach and intestine. The oil and butter mixture is buffed into two volumes of marijuana, parsley, lactose, or any edible inert powder, and then stuffed into large gelatin capsules.

## Seven

# Smoking Oil by Direct Vaporization

There are several customary methods for smoking of oil by direct vaporization with heat. The most common is the glass oil pipe, or vapor pipe. The oil is placed in the glass bowl of the pipe and the pipe is heated from below with a flame, similar to the method for smoking opium. This is a very efficient method, as very small amounts of oil may be vaporized at one time.

Another method for smoking oil straight is to place a tiny dab of it on a piece of aluminum foil. The foil is then heated from below with a match and the smoke is inhaled from above the boiling oil through a tube or funnel. This is essentially the same as smoking in a glass pipe, except that a new spot may be used each time and there is no build-up of residue at the point of vaporization. Before using, the foil is heated in a gas flame to burn off any part of the foil which might also vaporize.

# Eight

# Preparation of
# Translucent (Honey) Oil

One of the most potent and popular of the cannabis oil preparations is a thick, translucent, amber oil which has been extracted from Afghanistan hashish. This consistency is obtained by removing the colored impurities from cannabis oil that has been purified by re-extraction with petroleum ether.

The purified cannabis oil (which may or may not be isomerized or acetylated) is dissolved in ten times its volume of pure alcohol. An amount of granulated activated charcoal (Norit) equal to half the weight of the oil is added to the solution. The solution is filtered through fine filter paper and the alcohol is removed by evaporation. The residue is a translucent amber oil with the appearance and consistency of dark honey.

# Nine
# **Preparation of "Reefers"**

The term "reefer" has sometimes been used to describe a marijuana cigarette which has been impregnated with cannabis extract. This may be accomplished by working the cannabis oil in with the marijuana or tobacco to be rolled, painting the paper with oil before rolling, dipping the rolled joint in tincture of cannabis and letting it dry, or injecting the rolled joint with cannabis tincture and letting it dry.

# Ten
# High-Volume Extraction Method

The high potency and small volume of cannabis oil, in conjunction with anti-marijuana laws in many states, has added up to high potential profits for smugglers of the oil and new problems for law enforcement agencies. In the United States, the Drug Enforcement Administration (Drug Enforcement, 1973) has commented on the new problem (see Michael Starks' *Marijuana Potency* for an excerpt) and describes various devices they have seized. The plight of the DEA is reminiscent of the traditional moonshiner-revenuer chase, and no discussion of cannabis alchemy would be complete without a look at the hardware and methods of the large-scale operation.

The apparatus is constructed using two 55-gallon oil drums and equipment purchased from a hardware or surplus store, with which very large amounts of marijuana or hashish may be extracted and processed.

The apparatus is appropriately designed to meet the unique problems inherent in high-volume extraction and contains the necessary safety features to prevent mishaps with flammable solvents. The apparatus may be used to perform all the operations: refluxing, soxhleting, distilling, and collecting of solvents.

Even though the apparatus has many safety features (pressure relief valves and construction design which would prevent a minor mishap from being disastrous), the large volumes of inflammable solvents used pose a serious fire hazard. Even if the possibility of an accident is remote, the large scale and violence of a major accident warrant all possible caution. Such a device is always tended during operation, and measures to shut it down are ready at all times.

The solvent and cannabis solutions and mixtures are heated, via the tub of boiling water, in the lower oil drum. The upper drum acts as a giant condenser; it is filled with circulating icewater so that when the solvent fumes contact its surfaces, they recondense into liquid and fall back into the lower drum. The tops of the tubes, except for one which acts as a safety pressure relief valve, are closed with rubber stoppers. The solvent fumes condense inside these tubes, as they do on contact with the bottom surface of the condenser drum. The bottom surface of the condenser drum acts as the top surface of the lower boiling drum.

Construction of the condenser drum begins by marking corresponding four-inch grids on the top and bottom surfaces of the oil drum. A hole is drilled at each point where the lines intersect and pieces of one-inch copper tubing are run through the drum lengthwise. Each tube protrudes about one inch from the surface on both top and bottom. A watertight seal is soldered or welded around the outside of each tube where it passes through the surface.

Inlet and outlet fittings for the circulating icewater coolant are fitted to the top of the barrel. Two metal straps are attached to the top of the drum on opposing

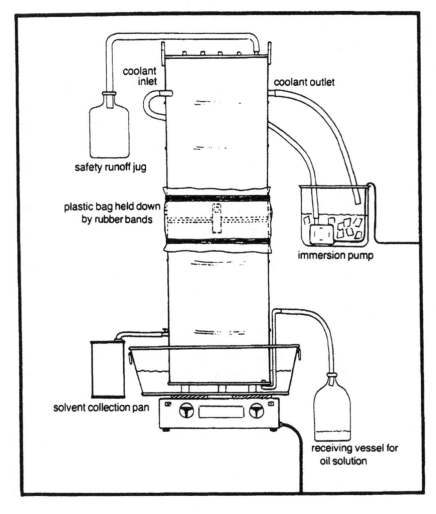

Figure 10.1

sides. The straps have large holes through which a chain is run. This allows the apparatus to be lifted with a small overhead crane.

The removal of the top is the first step in the preparation of the lower oil drum. Sturdy legs are attached to the bottom of the drum for maintaining a space between the bottom of the drum and the bottom of the boiling water bath into which the drum is set. A copper drain tube with an on/off valve is attached to the bottom of the drum, enabling the solvent/cannabis solution to be drained by siphon. Four metal straps are attached to the top of the drum. These protrude above the top of the drum and are used to secure the upper drum to the lower. A fitting with an on/off valve is run through the side of the drum approximately halfway between the top and bottom.

During extraction, a chamber containing marijuana or hashish is placed between the drops of recondensed solvent falling from the condenser and the boiling solvent solution. The distilled, recondensed solvent runs through the cannabis material, washing out the oil. The soxhlet chamber is removed and many holes are punctured in the bottom with a nail. A hole is cut in the side of the drum at a point slightly higher than the on/off valve in the side of the lower drum. A plate is fashioned to close the hole when necessary.

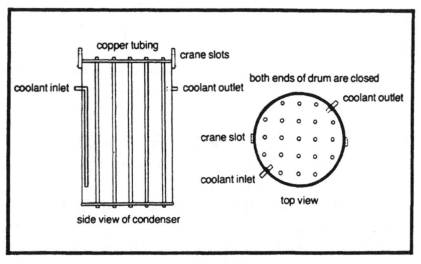

Figure 10.2

Solvent may be removed from a solution and collected by catching the recondensed drops of solvent in a funnel as they fall from the copper tubes. A tube leading from the funnel through the hole in the soxhlet chamber to the on/off valve in the side of the lower oil drum transfers the pure distilled solvent to the outside of the apparatus, where it is collected in a metal can.

The following additional equipment is used during operation:

1. A large, deep tub, at least twice the diameter of the oil drums, for a boiling water bath.

2. An immersion pump and a large pail for ice and water.

3. A long piece of hose which runs from the top of one of the copper tubes to an empty jug.

4. Rubber stoppers for closing the tops of the remaining copper tubes.

5. Large, thick polyethylene trash bags and several giant rubber bands, fashioned out of inner tubes, to fit around the drums.

6. A large funnel of nearly the same diameter as the large drum and a piece of tubing to connect the spout of the funnel to the on/off valve on the side of the lower drum.

7. An overhead chain winch or locking block and tackle for lifting the components.

8. Three heavy-duty sparkless electric hotplates.

The principles of each operation parallel those in the basic extraction method using kitchen and hardware equipment.

The marijuana or hashish is prepared for refluxing by grinding it to a fine powder. The lower drum is

filled to approximately one-third full with the powdered cannabis material and solvent is added to half
fill the lower drum. The valve on the side of the drum
is closed. The funnel and soxhlet chamber are not yet
used. The upper drum is secured to the lower and the

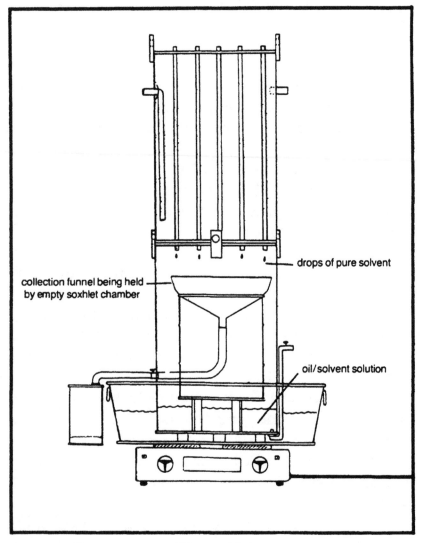

Figure 10.3

crack between the two oil drums is sealed in the following manner: a large trash bag, with the bottom cut out, is slipped over the apparatus and positioned at the joining of the drums. Inner-tube rubber bands secure the plastic to the drums both above and below the joining. This arrangement acts as a safety valve. Should any pressure build inside the apparatus, the plastic will puff out.

The entire apparatus is placed in the tub, which is resting on the three electric hotplates. An immersion pump in a tub of ice and water supplies coolant, which is pumped through the condenser drum. Rubber stoppers are used to seal the tops of all the copper tubes except one. A long piece of garden hose is attached to the tube; the hose runs into a large jug which is set in a place which is safe from flame or electrical spark. As long as icewater is being circulated through the condenser drum, no fumes or liquid will appear in the jug. If any fumes or liquid are given off from the hose, the boiling water is removed from the bath, and ice and water are immediately added. The condenser and coolant system are then checked.

The electric hotplates are turned on and allowed to heat the water bath to boiling temperature. This, in turn, heats the mixture of solvent and powdered cannabis material. The solvent fumes rise to where they contact the copper tubes and bottom surface of the condenser barrel, then recondense into liquid and fall back into the boiling mixture.

The mixture is refluxed for three hours, then the hotplates are turned off and disconnected. The boiling water is removed from the tub. The tub is refilled with ice and water, and allowed to stand for at least

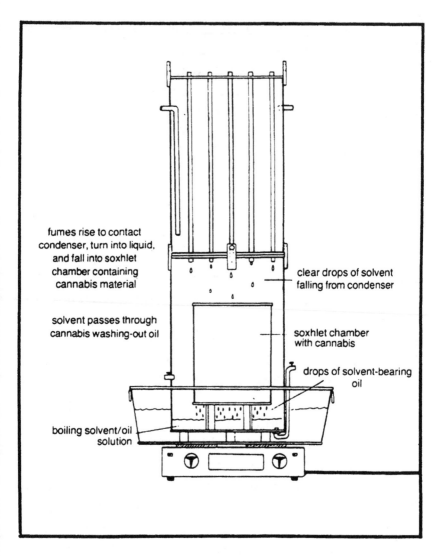

fumes rise to contact
condenser, turn into liquid,
and fall into soxhlet
chamber containing
cannabis material

clear drops of solvent
falling from condenser

solvent passes through
cannabis washing-out oil

soxhlet chamber
with cannabis

drops of solvent-bearing
oil

boiling solvent/oil
solution

Figure 10.4

one-half hour. After the apparatus has cooled sufficiently, the oil-saturated solvent is removed using a
siphon pump to draw the mixture through the drain
tube. After removing all the solvent and draining the
condenser, the drums are opened, and all the cannabis
material is scooped out and stored in a closed drum.

The cannabis oil/solvent mixture is replaced in the lower drum and assembled for distillation and collection of the solvent. The funnel, which is held by the empty soxhlet chamber, collects the recondensed drops of pure solvent as they fall from the tubes. The liquid runs through a hose to the fitting on the side of the drum. Another hose on the outside takes the solvent to a receiving barrel, where it is collected. When all the solvent is evaporated, a heavy film of oil coats the bottom of the lower drum. This is redissolved in a small amount of solvent and the mixture is removed from the drum. The mixture is stored in an unbreakable container.

The damp, powdered cannabis is replaced in the drum and the solvent just removed and collected is added. The apparatus is assembled and operated as before, then refluxing for three hours, draining and collecting the oil.

The bottom of the soxhlet chamber is fitted with a filter paper and set in the lower drum. The chamber is filled with the powdered cannabis, and clean solvent is added to the cannabis until it is saturated and several inches of solvent, which have run through the cannabis material, are in the bottom of the drum. The apparatus is reassembled and the water in the bath is brought to a boil. The fumes of the solvent in the lower drum will rise, recondense to liquid in the copper tubes, and fall into and run through the cannabis, washing out the remaining cannabis oil. After several hours of soxhleting, all the oil will be dissolved in the solvent. The cannabis oil/solvent solution is distilled and collected as before. The oil remaining in the drum is dissolved in alcohol and removed. This solution is added to those collected after each refluxing and combined in a large metal pot, the weight of which has been noted. The pot is placed in the dry bottom of the lower drum and the apparatus is assembled for distillation and collection of the solvent. The solvent is removed and the product remains in the pan. If further purification and chemical alteration is desired, the methods given earlier are applied.

Eleven
# Advanced Refinement Techniques

The translucent amber oil produced by charcoal-filtering the ether phase of the extraction and isomerizing the cannabidiol present to THC contains, in most cases, between thirty and sixty per cent THC. Utilizing rather complex and exacting techniques of modern chemistry, it is possible to further refine this oil. Fractional distillation of the oil will yield a product which is up to twice as strong as the ether phase, and can be converted into nearly pure THC. Totally pure THC, a thin transparent oil, can be produced by chemically isolating the pure cannabidiol and then isomerizing it to THC. This is a very complex chemical operation and requires much sophisticated equipment and chemicals. These advanced laboratory techniques are probably beyond the reach of the starting alchemist but are important in the sense that they lead to the production of the THC of highest purity.

## Fractional Distillation

Fractional distillation of the oil requires that the oil be heated to a high temperature under a reduced pressure created by a vacuum pump. This causes the THC and related cannabinoid substances to vaporize. The vapors are condensed back into an oil on contact with a cooled surface. The desired fraction is collected by selecting the appropriate temperature and pressure for the distillation. Many of the impurities do not vaporize and are left behind in the flask used for heating the oil. The following is a description of a laboratory method for refinement of crude red oil and purified red oil from the basic extract. The work was done by Roger Adams in 1940 and appears on page 198 of volume 62 of the *Journal of the American Chemistry Society*.

Wild hemp, grown in Minnesota during the season of 1938, was used in the following experiments. The hemp plants were cut after flowering had begun but before seed had set in the female tops; they were stored in a room for six weeks to dry out. A fan was used for circulation and no molding was evident. One-third of the dry hemp plants amounted to stems. These were held and shaken to remove the leafy part of the plant. This clean marijuana was extracted with 95% pure ethyl alcohol. The methods of extraction are described below.

Four twenty-gallon crocks, each with a capacity of 23 pounds of material, were arranged for countercurrent extraction. Each crock held 61 liters of solvent, of which 40 were withdrawn at each transfer, with 20 liters being retained by the cannabis. After the process had become uniform, the extract of crock #4 at each transfer held approximately 2 gm of solids per 100 cc. Transfers were made once or twice a day as necessary. The most concentrated extract obtained in this manner was passed to a concentrater, where most of the solvent was flashed off under vacuum. Never was the temperature raised above 50°C. The evaporation was carried out at 30°C. The concentrated solution contained 23.1 gm of solids per 100 cc 95% ethanol, and each 1 cc represented 4.13 gm of hemp.

The red oil from these extracts was obtained by the following methods: ethanolic extract was poured into a 1-liter Claisen flask with a short, wide neck and filled with glass wool until the flask was two-thirds full. The temperature of the bath was raised gradually from 90° to 140°C as the pressure was diminished slightly. The distilled ethanol was discarded, and the flask was again filled to two-thirds capacity. This process was repeated until 1600 cc of extract had been added and the alcohol removed. The temperature was then raised to 200°C, and when the last traces of ethanol ceased, the bath was lowered to 180°C and the pressure reduced to 30 mm. Care was necessary to prevent the liquid from foaming over. The temperature was raised gradually to 200°C until distillation ceased.

The bath was then cooled to 170°C and the pressure reduced to 2–5 mm. The residual product was then distilled. Much care was necessary to keep the bath at the lowest temperature at which the oil distilled evenly, since there was a marked tendency to foam. The material distilled at between 100° and 220°C (3 mm) with the bath temperature at 170–310°C. Yield 180–200 gm crude red oil.

This product was dissolved in 500 cc 30–60°C b.p. petroleum ether and extracted several times with water. The ether layer was distilled and the residue fractionated through a good column having an outside heating unit. The first fraction boiled at 115–120°C and gave a yield of 70–80 gm. The second fraction distilled at 150–175°C, yielding 25–30 gm. The material remaining in the flask was removed by dissolving in ethanol and filtering from the glass wool. The ethanol was evaporated and the product distilled from a 250 cc flask, b.p. 175–195°C (2 mm). Bath temperature was 220–270°C. Yield 90–110 gm purified red oil.

# Chromatography

An advanced separation method known as chromatography may be used to remove non-active elements from the product. This involves filling a tube with a material which retains unwanted constituents of the oil. The oil is dissolved in a solvent and passed through the material. Chromatography of the hexane extract of hashish in the following formula (hexane is a solvent with properties similar to petroleum ether) removed unwanted constituents of the oil amounting to 49% of its weight. The chromatographed extract obtained was almost totally composed of cannabinoid elements. After conversion of the non-active elements of the oil, the resultant extract will be nearly all THC. The process following is derived from the *Lloydia Journal of Natural Products*, page 456, vol. 33, no. 4.

**Isolating the cannabinoids from hashish**

The National Institute of Mental Health supplied 13 kg of confiscated hashish, origin unknown. The hashish was extracted in a stainless-steel pot, using

95% ethyl alcohol at 50°C, and was stirred for five hours. A second and third extraction were then completed using 32 liters for 24 hours and 20 liters for 72 hours, respectively. The combined hexane hexane extracts were washed with 5 liters of 50% aqueous ethanol. The solvent was then removed in vacuum at 40°C to provide a 22.9% recovery, or 3056 gm. This is shown by gas-liquid chromatography to contain 29.5% cannabidiol, 8.2% cannabinol, and 5.8% $\Delta^9$-THC. Florisil (30.5 kg) and methanol (2%) in hexane were used to chromatograph this oil. The resulting dark oil contained 50% cannabidiol, 20% cannabinol, 15% $\Delta^9$-THC, and 15% unidentified components. Although using Florisil (40:1) provided essentially pure cannabidiol by gas-liquid chromatography, the product could not be induced to crystallize. Crystallized cannabidiol is obtained by using the following modified procedure of Roger Adams.

**Isolation of pure cannabidiol**

If completely clear THC (a clear, thin, colorless oil) is desired, it is necessary first to isolate pure cannabidiol from the chromatographed oil by converting it to cannabidiol-bis-3,5-dinitrobenzoate. This is then converted back into pure cannabidiol, which is now in the form of white crystalline prisms. The process for this operation is found on pages 456 and 457 of the *Lloydia* volume previously mentioned, and a description of it follows.

Cannabidiol-bis-3,5-dinitrobenzoate is made by rapidly adding 300 gm fresh 3,5-dinitrobenzoyl chloride (m.p. 68–69°C) to a mechanically stirred solution of a chromatographed hashish extract in dry pyridine at 0° under nitrogen. The mixture was stirred for 15 minutes, then warmed in a 60°C hot water bath for 30 minutes. This mixture was then poured into a mixture of 200 gm of ice and 300 ml concentrated hydrochloric acid and extracted with ethyl acetate (750 ml). The precipitate was filtered and washed with another 750 ml ethyl acetate. The aqueous phase was separated and washed with 500 ml ethyl acetate. The combined organic phases were washed with aqueous sodium bicarbonate (2 × 200 ml) followed by 300 ml distilled water and dried over $CaSO_4$. The solvent was removed in vacuum to yield 340 gm of a dark oil. This was purified by crystallization from 1800 ml ethyl ether, yielding 194 gm of off-white powdered cannabidiol-bis-3,5-dinitrobenzoate melting at 97–101°C.

Pure cannabidiol is made by adding 220 ml of liquid ammonia to a solution of 288 gm cannabidiol-bis-3,5-dinitrobenzoate in anhydrous toluene (400 ml) at −70°C in a Parr bomb. The sealed apparatus was mechanically stirred. During five hours the pressure built to 110 psi and the temperature rose to 20°C. The ammonia fumes were released overnight. The product was dissolved in heptane (400 ml) and insoluble 3,5-dinitrobenzamide was removed by filtration. The precipate was washed twice with 150 ml heptane. The heptane solutions were combined and washed with boiling water (5 × 200 ml) and the solvent removed in vacuum to yield 120 gm of a dark oil. Chromatography on 180 gm of this product on 3400 gm of Florisil and elution with 30% chloroform in hexane yielded oily cannabidiol (140 gm). Crystallization from 30–60° petroleum ether yielded 99.2 gm white prisms, and recrystallization gave 94.8 gm pure cannabidiol.

# Conversion of Pure
# Cannabidiol to Pure THC

The crystalline prisms of cannabidiol are converted to pure THC utilizing a formula of Roger Adams found on page 2211 of volume 63 of the *Journal of the American Chemistry Society*. The following is a description of a method for producing pure THC.

## Isomerizing the cannabidiol with sulfuric acid

One drop of 100% sulfuric acid was added to a mixture of 1.94 gm crystalline cannabidiol in 35 cc cyclohexane. After refluxing for one hour, the alkaline beam test was negative. The solution was decanted from the sulfuric acid, then was washed twice with aqueous 5% bicarbonate solution and twice with water. It was then evaporated. This residue was distilled under reduced pressure to yield pure THC with a rotation range of 259° to 269°.

# Appendix A

# Letters

D. Gold invented a machine called the Isomerizer, followed by the Iso II. The following letters were received by Thai Power, the manufacturer of the machine.

## CYCLING EXTRACTION TECHNIQUE

*The following is a letter written to us from an alchemist who has researched several methods of speeding up the pre-isomerization extraction process. (Editor's Note: Thai Power encourages experimentation but in no way can be responsible for accidents resulting from any deviation from the instructions.)*

Dear Thai Power:

As all users of the Isomerizer® know, before the cannabis is isomerized, the oil must first be extracted from the grass or hash. The Isomerizer® instructions tell how to do this by a soaking and soxhleting method. This is very efficient and no doubt the best general method of extraction. There are, however, several variations I have tried which have proven useful under certain situations.

As you folks are keeping in contact with your public, I thought you would want to relay my findings. I have found that soxhleting time and actual machine operation time can be cut considerably if you increase the soaking time by following this procedure:

1. Soak for 24 hours as indicated in the instructions.

2. Pour the cannabis/solvent mixture through a coffee filter. Save the green solvent and replace the wet herb in the soaking vessel.

3. Add the fresh solvent to the herb in the soaking vessel and soak for another 24 hours.

4. Pour off the solvent through the filter paper, add the green solvent to that obtained before, and resoak again in fresh solvent. Repeating this process several more times has cut my soxhleting time to often less than two hours for complete extraction. I work during the week and I like to have as much time as possible on the weekends, so I have used this process to cut my actual machine operation time to just a few hours. I begin my soaking on Sunday and change the solvent every night until the next Saturday. This takes about five minutes each night. I begin soxhleting Saturday morning, and after a very short time the cannabis is almost completely brown, indicating a good extraction. I then Isomerize® and am smokin' by noon. It seems to me that if you have cannabis and solvent you should always have some soaking.

Sincerely,
Steve
Waukegan, Illinois

*Dear Steve,*

*Several extra cycles of soaking in fresh alcohol will cut the extraction time considerably. It should be mentioned that each time you pour off and save the green solvent, the total amount of solvent which contains the oil increases. These combined extracts, when poured into the reaction vessel, should not fill it to a level above one-half inch below the soxhlet basket. If there is too much solvent/oil solution, simply operate the machine in the normal manner for removal of solvent and concentrate the solution down to the prescribed several-inch level.*

*The following letter was received from an experimenter who has been successfully using petroleum ether.*

Dear Thai Power:

I have been using petroleum ether (30 to 60° boiling point) to refine my isomerized oil and would like to know if you think my method is safe and efficient. Here's what I do:

After soxhleting and neutralizing, I remove the solvent until there is about 6 ounces of green solvent/ cannabis oil solution remaining. I pour this into a quart-sized Pepsi bottle, half filled with water. After the solution has cooled to room temperature, I add enough pet ether to raise the level in the bottle by one inch. I put the screw top back on, invert the bottle slowly, return it to its upright position and release the pressure by loosening the screw cap. I do this 15 times and then let the bottle stand until the liquid separates into layers.

The top layer contains purified oil dissolved in the pet ether. I siphon this off using a large rubber squeeze bulb attached to a ¼″ flexible tube. The pet ether is now gold in color. All the green stuff seems to stay in the alcohol-water solution. I add more pet ether and repeat the process until the pet ether remains clear after the 15 inversions.

I combine the gold-colored pet ether extracts and pour them into a large, flat-bottomed Pyrex pie plate, which I put out on the roof on a hot, sunny day. After several hours all that remains is a translucent gold oil of unbelievable potency.

What do you think?

Name Withheld

*Dear Extractor:*

*You are using an efficient method for removing some of the non-psychoactive organic impurities from your oils. As to safety, your method is certainly much safer than attempting to remove petroleum ether in any manner on a stove or hot plate. In an operation such as this, a general rule would be — the less the volume of petroleum ether used, the safer the operation! Other than that the safety depends on the amount of care you take with cigarettes, sparks, pilot lights, etc.*

*Thai Power is in the process of bringing out an automatic filtration accessory to the Isomerizer® which offers a much finer degree of purification with greater ease and safety than your petroleum ether method.*

# TEXAS SUPER HASH

Dear Sirs:

I have recently discovered several excellent hash-making methods using the Isomerizer®.

"2-on-1 Hash" is made from regular Mexican, costs me about $15 per ounce, and is as good as any imported hash I have ever smoked. I call it "2-on-1" because two ounces of weed are extracted, isomerized, then the oil is put on one ounce of isomerized weed which has been ground to a fine powder. I have found a very simple, fast way of making this using the Isomerizer®. Here's how I make one ounce:

1. Take three ounces of commercial Mexican weed, clean it through a strainer, spread it ¼" thick on a cookie sheet and put it into an oven which has been preheated to 250 degrees then turned off. After 15 minutes take it out and crumble it between your thumb and forefinger. If all the water is removed the weed will crumble easily into powder. Take ¾ ounce (21 gm) of the dried weed, put it into a blender and add four ounces of alcohol. Run the blender for five minutes at the highest speed, reducing the cannabis to very fine particles suspended in the alcohol. Pour this mixture directly into the reaction vessel.

2. Take the remaining 2¼ ounces of weed and put it into the blender with eight ounces of alcohol. Run the blender at high speed for ten minutes.

3. Fit a large, chemist-type coffee filter into the soxhlet basket and place it into the reaction vessel. *(Ed. Note: The reaction vessel already contains the original ¾ ounce of weed in alcohol.)*

4. Pour the alcohol and 2¼ ounces of weed in the blender into the soxhlet basket. After all the alcohol runs through the filter paper you have ¾ ounce of weed and the alcohol in the bottom of the reaction vessel, and 2¼ ounces of weed in the filter paper held by the soxhlet basket.

5. Put the reaction vessel in the machine and run it as per directions for two hours. The weed is ground so fine that this seems to be plenty of time to extract all the oils. (*Ed. Note: The finer the cannabis is ground, the shorter the extraction time. We estimate this method would remove about 80% of the oil from the cannabis contained in the soxhlet basket.*)

6. After the two hours, add 15 drops of concentrated activator solution to the weed and alcohol in the bottom of the reaction vessel. Replace the soxhlet basket containing the 2¼ ounces in the reaction vessel and leave it in during the Isomerization® to wash out any residual traces of oil left in the 2¼ ounces.

7. After 45 minutes open the solvent removal valve and start removing the solvent. After about 15 minutes, all the solvent is gone.

8. Open the machine, neutralize, remove the soxhlet basket and discard the 2¼ ounces of leached weed.

9. Put the reaction vessel back in the machine, turn on the pump and dry the weed as per instructions. It dries very fast since it is so finely powdered.

10. After the weed is dry, scrape the reaction vessel with a plastic spatula and put the oily powder into a large mortar, then stir with a pestle. After it is.mixed completely, add two or three grams of water and work it evenly into the cannabis. The water is very important because it contributes to the taste, smell and

appearance of the hash. Voila! One ounce of really great hash!

P.S. I tried this with some Colombian red and it was just plain too damn strong!

Nobody could smoke it and stay conscious. I did find that if Colombian is prepared in the blender and isomerized in the normal manner, a fine, deep-brown, earthy Kif-like smoke results which is great to smoke in a pipe. Adding a little water is very important here also.

I mainly smoke a bong. I believe that because of its deep-breathing and total-burning aspects, a bong is maybe twice as efficient a method of smoking as joints or a regular pipe. (*Ed. Note: We heartily agree. Isomerizing your cannabis and smoking it in a bong will yield many times the enjoyment per ounce.*) The water makes the Iso-Kif bind together when lighting the bong and also cools the smoke. I tried using "lettuce opium" as a binder in my various hashmaking experiments. It smoked nicely and may have even helped the high. That is hard to tell though, considering the intensity of the Iso-Buzz.

Sincerely,<br>
Tex<br>
Austin

*Editor's Note:*

*Tex has come up with a good method for making hash. Especially considering those $60 pounds available in Texas. There are several aspects of his process we would like to comment on, however: The two-hour extraction time probably won't get all the oil from the cannabis in the extraction basket. Overnight soaking* *should always be the general policy, and if the cannabis is blended with alcohol, the soaking vessel should be shaken lightly every few hours. Also, Tex removes some of his alcohol during the isomerization process before neutralizing. This is OK to do, but all the alcohol should never be removed before neutralization. If this should happen, be sure to add alcohol to the reaction vessel before adding neutralizer.*

## LIFETIME STASH

Dear Sirs:

I am now growing a modest-sized plot of high-quality sinsemilla for my own consumption. My small plot (only 40 × 40 feet) yields one heck of a lot of grass. I was pleased with the potency increase due to Isomerization and even happier with the fact that the high is different and more pleasurable than any non-Iso® cannabis. I was kind of pissed, or rather frustrated, that I could only process 4–5 ounces at a time and that the operation took 7–8 hours to run, and I had so much weed that I could never hope to process it all.

Then I received the first issue of IsoNews®. Hooray for Tex and the blender!! I purchased a used industrial-commercial food preparation blender from a restaurant supply house. It cost just under $100—and looked just like your standard Waring except that it was 4 times as big and 10 times as strong. I bought this blender after burning out two kitchen Waring blenders.

I found that after the weed was thoroughly dried I could easily reduce several pounds of flowers and leaves to fine powder in just a few minutes. The larger stems were removed, and since I had carefully destroyed all the male plants, there were no seeds — sinsemilla.

This blending process reduced the volume about 90%. I basically used Tex's Isohash method except that I put 1½ lbs. of pulverized leaf and flowers in the soxhlet basket and 4 ozs. of powdered flowers in the alcohol/oil solution. I had to use a big filter paper and pack it to the top.

It seemed like a good idea to keep as much alcohol in the RV as possible. I always kept the RV full to within ¾" of the basket. Was this necessary to facilitate Isomerization or extraction? It took only 3½ hours to extract the 1½ lbs. in the basket. After Isomerization, neutralization and solvent removal I had 5–6 ozs. of psychedelically overwhelming Iso®hash. Even the smallest little piece (the size of a matchhead) of this gooey black hash will, when smoked in a glass oil pipe, keep me psychedelically stoned all day.

For that special Sunday high, I made a batch where I added 2 ozs. of dried and powdered homegrown *Psilocybe cubensis* mushrooms to the powdered grass in the soxhlet basket. A little dab will really do ya!

I figure that by the end of harvest time this year, I should have a block of 8-on-1 Isohash® stashed away in my deep freeze big enough to last my wife and me the rest of our lives. Not a bad deal for one summer's work and a $500 investment!

I am concerned that my block of hash will retain its present potency for a long time to come. I specifically made it into gooey, resinous Iso®hash because I remember reading somewhere in the scientific literature that cannabis extracts retain their potency much longer than the raw cannabis. This seems logical, as the oils on the outside of the lump hermetically seal off all the cannabinoids.

When I'm done harvesting, I figure that I should have a cube of this stuff about 1 foot on each side. Is the best way to store it to wrap it in several layers of lightproof plastic and then ice it away in the bottom of my deep freeze, only opening the package when I need another ounce?

Thanks again for the Iso®News, with special thanks to Tex.

K.O.

*Dear K.O.:*

*Congratulations on your lifetime stash. A truly enviable position.*

*To answer your first question, you were right to use the maximum amount of alcohol when processing a relatively large amount of grass, especially when the powdered cannabis is left in the alcohol/oil/acid solution. As a general rule, whenever processing over several ounces of cannabis, especially when not powdered, fill the reaction vessel to a depth of at least several inches.*

*You're right again on the long-term storage of cannabis. There is probably no better form than 8-on-1 Isohash for long-term storage. What happens when cannabis gets old and loses its power is that the THC oxidates into a non-psychoactive substance. Exposure to air, light and heat cause this to happen. Any substance that has been processed should be kept in an airtight, lightproof container in the freezer.*

# BUNK TO BEST

Dear Thai,

Hi. I just came up with a hash-making method that produces a product far superior to that of Ol' Tex! I just happened to find this out while checking out that article on the Isomerizer® in *Head* magazine. I was familiar with the Isomerization® process as I had gone to a college medical library and looked up the formula reference given by David Hoye in *Cannabis Alchemy*. Kicking the cannabidiol element over to high-rotating THC is a pretty clear-cut thing. Well, I decided to dig into this decarboxalation trip and see what it was all about. It was while checking this out at the library that I came across the hash-making method that should cause me to go down in the annals of Hash-Making History. I kid you not, Jack, I can make hash out of good regular Mex that any Colombian, Thai, high-altitude snob has to admit is better than his shit at a minute portion of the price. This stuff is far higher quality and much more pleasurable to taste, smell and smoke than that greasy, black, tarry Texas Isohash.®

The way I make this stuff is with what I call the Pittsburgh Doublewash® system.

This marvelous hash-making system was born when I found a scientific paper in one of the chemical journals that discussed the whole decarboxalation trip, but at the same time said that if you make marijuana tea, boiling the shit out of it in lots of water for one hour, you can throw away the tea, dry out the marijuana, and it will be considerably better than before. This pisses me off a bit because I have been drinking grass tea for years and throwing away the weed. Anyhow, I can see how the oils can be insoluble in water, unlike some of the tars and waxes, but I would sure think that all that violent boiling would loosen up the oils and mix them with the water. But they had analysis graphs that showed this was not the case. The boiling water extraction just removed about 20% by weight of the water-soluble tars and waxes. Simply boiling in a pan of water for 1 hour also decarboxalated all the THC acid to THC. From what the literature states, the Isomerizer® also accomplishes this during the extraction process. I did several tests with boiling water and decarbox and non-decarbox (I didn't Isomerize either). The boiled batch smoked with almost the same potency as the non-boiled. The boiled was much stronger than the non-decarbox when eaten. This indicates that the THC acids decarbox fairly thoroughly on smoking, and the only real application of straight decarboxalation is when you are planning to scarf the product.

But the big discovery was yet to come.

I took the stuff that had been boiled for an hour, poured it through a T-shirt filter and drank the tea. Not much happened except for a heavy sensation in the head and limbs. I added fresh distilled water to the pulverized weed and boiled it for 3 more hours. Again I poured it through a T-shirt but this time let the brown sauce go down the drain. I'm convinced that there was no dope in it. I squeezed the weed, now much lighter in color, nearly dry in a wine press and did the final dry in a turned-off 300° oven.

This constitutes Part 1 of the Pittsburgh Doublewash® method. In Part 2, I take this stuff and, as if it was any other starting material, chuck it into the blender with some isopropyl. I then extract and Isomerize.® I always leave all the weed in the soxhlet basket. Tex's method of leaving some ground-up pot floating around in the bottom might somehow inhibit the full Isomerization process. The same may be true with some of those tars and waxes that aren't in there because I washed them away in the first water wash. Anyhow, I Isomerize, neutralize, and then dump all the weed back into the reaction mixture. After I pump off the alcohol I wash it again with water to get rid of all that harsh, left-over baking soda. After this wash, I dry the pulverized pot again, put it back into the soxhlet basket and re-extract the Isomerized oil. After the weed is all extracted I dump about 1/5 of the leached powder back into the sauce and evaporate the alcohol.

I'll tell you what, that stuff tastes, smells and smokes even more pleasurably than most of the high-cost imported hashish. As I make about 2 ounces from each pound of good regular, each ounce costs me $40–50. This ain't at all bad considering that this stuff is superior by far to the Afghani Mazari-Bizarri Primo $150-per-ounce hash that is my alternative.

Up until I discovered my Pittsburgh Doublewash® system, all your machine could do was make bad grass better, good grass great and great grass greater, etc. Also real strong but yucko-tasting hash from regular Mexican. I, for one, could not make something more pleasurable than Thai Sticks or high-altitude Mich or Oaxacan out of some gross Tijuana bunk. But I now realize that this was due to all those rotten water-soluble tars and waxes. After Iso,® the THC is the same as the good stuff. All the bad-tasting headache shit is in those tars. This refinement is great for the lungs, too. I'm a chronic cougher when smoking regular hash or non-wash Isohash. After getting all those non-psychoactive tars out of there, I was able to nail three big hits in succession out of a toker pipe with only a minor cough.

Anyhow, using my new method I can finally say that it is possible to make something that will satisfy even the $150-per-ounce super smoke snob out of regular old commercial for under $30 per ounce.

Name Withheld
Pittsburgh, PA

*Editor's Note:*

*Congratulations, what more can we say.*

## DOUBLEWASH ISO OIL THE BEST

Dear Thai Power,

For the past few months I have been making and comparing all the different Iso products that I knew of. What I felt to be the most important and valuable aspect of smoking is the overall pleasurable aspects yielded by the actual smoking of the cannabis. By this I don't mean just strength, but potency combined with taste, ease in smoking, effect on the lungs, immediate effects, aftereffects, whether or not it gives you a roaring case of munchies, etc. And, of course, all this must be weighed against cost.

What I actually have been doing is determining how to get the ultimate pleasure for the least amount of bucks. Here are some notes on my experiments.

## Iso Hash

I made several different batches of Iso hash from several different varieties of weed. I employed the doublewash technique on most of the regular weed I processed. I found that the doublewash technique greatly improved the taste and smell of the Iso hash made from musty-smelling and harsh-tasting regular. It seemed that the amount of actual pleasure derived from the Iso hash was determined by the actual amount of oils used in ratio to the amount of powdered material. I found that the best hash I made had a 35–40% oil content and was approximately the same consistency as fresh powerful Afghani patties. I made some 3-on-1 Iso hash from some Columbian red buds. As I love the spicy flavor of Columbian I omitted the hot water pre-wash and only used the second cold water wash to remove the leftover baking soda. This was without doubt the strongest hash I've ever made or smoked. Pleasurewise it was ecstatic if smoked just before going to sleep—because that's what you are going to do soon after smoking it any time. The big problem with this stuff is that it costs a small fortune to make. After cleaning the Columbian buds, I had less than half left, so in order to get the three ounces that I concentrated into one ounce, I had to clean 8 ounces of buds.

## Concentrated Iso Weed

My partner in these experiments is what you might call a paper addict. Even though he realizes the wastefulness of smoking joints (as opposed to a bong or a "smokeless" pipe), he likes to smoke joints. About the most pleasurable rolling preparation we made was some 1-on-1 Columbian red. The Isomerization changed the high to a very awakening and crystal-clear high. The cost was way up there still, as the Columbian gave a 40% yield after cleaning. What we found to be the best deal in rollable preparations was 4-on-1 made from fairly good green Mexican regular with the doublewash process. The trick to keep it cool-burning and good-tasting is to leave just a small amount of water in the cannabis from the final washing. We very successfully kept this in an airtight bag in the freezer. Only the amount we were about to smoke was thawed. This was easily done by holding the weed between the palms for several seconds.

## Doublewash Oil

Many of my friends don't like oil because they find it a royal pain in the ass to handle and smoke. This is true if improper techniques are used. Using a wooden match to dip oil from a bottle and then painting the oil on a paper prior to rolling can get very messy (also very wasteful, as the oil-impregnated paper is directly exposed to the air and much of the oil smoke is lost). Glass oil pipes always give you a solid hit, but after they are used several times they get that black carbon all over them which in turn gets all over everything—along with the oil. Also, the hit from the glass pipe is very apt to give you a ten-minute choking and coughing fit.

The best way to prepare oil for smoking with weed is to work it into a pile of weed with the fingers. This evenly distributes the oil into the weed but stains the hand and whatever it touches for days.

Well, I have come up with the very best way to smoke oil efficiently with no hassle, and I have found that doublewash oil from good regular smoked with my method absolutely gives the most pleasure for my smoking dollar. I get about 2–3 grams from each ounce of regular weed that I wash and extract, which means that the DW oil is costing me $4 or less per gram. When this oil is smoked using my new method, there is no throat or lung irritation and absolutely no tendency to cough. The taste is sweet and rich, and I find that I like this taste as much as the taste of the super-expensive brands of pot.

The key to my smoking method is cigarette ashes. I have the large bowl of either a bong or a toker pipe filled with them. I use a 6-inch piece of coat hanger as a dipstick. This is put into the oil vial and a hit's worth is picked up on the end. I hold this directly over the ash-filled pipe bowl and heat the coat hanger with a butane lighter. This causes the drop of oil to fall onto the ashes and spread out. Then I smoke the pipe just like it was full of weed or hash. The ashes are totally inert as they have been completely burned once and all that comes through is the water-cooled taste of the DW oil. The process (of getting the oil from the bottle to the bowl with no hassle) is very easily repeated with little mess or waste. This is what I found to be the product that I like the best. Good-tasting, high-potency oil with the crystal-clear Isomerized effect just can't be beat at under $4 per gram.

Anonymous

*Dear Anonymous,*

*This is indeed a good way to smoke oil. Spreading the oil on an inert medium is very efficient. The same effect* as given from the ashes can be had by filling the pipe with steel wool or layers of cloth mesh pipe screens. A well-used and resinous pipe screen also works well. There may be a difference in the potency of the oil when smoked in this manner (placing the flame directly on the oil). When the oil is smoked in an oil vaporization pipe or on aluminum foil, the oil is not actually burned, but distilled. The heat of the lighter causes the oil to vaporize and your lungs act as a condenser. The product actually entering your lungs is slightly different than when the oil is actually burned. Perhaps this accounts for the more predominant tendency to cough when the oil is smoked in the oil pipe.

# BINDING & PRESSING ISO HASH

Thanks for the recipe for "Texas Super Hash", but can anyone suggest a good binding agent or some type of press for the hash? It seems that mine never hardens fast enough before I smoke it; and I'd like to add that people around here are amazed at what my little toy can do. Many thanks for introducing the Isomerizer to Massachusetts.

For a freer and greener Earth,
Joe

This letter comes from Joe in Massachusetts and inquires on a subject that many others have also asked about. When Iso hash is prepared with an oil-to-powdered-material ratio of less than 20%, the material does not bind together and is more the consistency of dark kif. As most hash, especially that from the Middle and Far East, has traditionally come as blocks and lumps (*due to the high resin ratio*), *many people have sought for an additive to cause Iso hash with a lower oil content to bind together. The best binder yet discovered seems to be an extract of lettuce made by drying the lettuce and then extracting it in the Isomerizer. After the solvent is evaporated, the tarry extract is scraped from the reaction vessel. A small amount of the extract is added to the Iso hash when it is worked with the thumb of one hand in the palm of the other. Be careful not to overdo the amount of lettuce extract used. A small amount will cause the cannabis to bind well when kneaded and worked. An excellent way to work the hash is to totally dry the Iso hash into powder and then put several teaspoons into the corner of a heavy plastic bag and add several drops of water and a small amount of lettuce extract. Squeeze and press the bag until the mass begins to adhere to itself. Soon it will form into a solid mass and is removed from the plastic bag.*

*Hash presses are another subject which has been asked about in many letters. There are several methods of pressing the powder into solid slabs that can easily be done at home. Always remember that a small amount of water added to the mass causes it to bind better.*

*Binding hash by compressing it tightly is a very good preservative measure because it hermetically seals all the hash within the outer crust. Chemical experiments done many years ago on a lump of Indian* charras *of about 100 lbs. weight showed that the crust deteriorated in a few years time to the point that it was 1/20th the potency of the protected hash in the center. Wrapping the hash tightly in light and airproof plastic and then freezing it is the best preservative measure that can be taken. A small piece is quickly thawed for smoking by holding it in the closed hand.*

# A NEW ISO OIL DEVICE

Dear Sirs:

I have come up with a good method for smoking hash oil in case your readers may be interested. I take a glass mason jar and drill a hole through the bottom. Then I take a piece of aluminum foil and place some oil on it. I then wrap the foil across the opening of the jar with the hash oil on the inside. Then I invert the jar and hold a flame under the foil in the spot where the hash oil is at. When the jar is filled wih smoke, inhale through the hole in the bottom of the jar.

Also I would like to say that I am well satisfied with the results of the Isomerizer.

Sincerely,
A Loaded Floridian

# LETTUCE OPIUM

Dear Thai Power,

Is the procedure for making lettuce "opium" any different than a normal oil extraction? Also, what is the best way to smoke the stuff?

Sincerely,
Art Smith

*Dear Art,*

*The dried, tarry extract of lettuce has currently become very popular in the various magazines. Perhaps this is due to the fact that a single head of lettuce can yield up to half an ounce of the preparations that are* *being retailed for up to $5.00 per gram. You don't have to be Albert Einstein to figure out this math.*

*A high-quality extract of lettuce can be made for pennies per gram in the Isomerizer or ISO-2 by following this procedure:*

*First, chop the lettuce as if you were making a salad. Dry the lettuce by spreading it out in the sun for several days or put it into a 250° oven until dry. Pulverize the lettuce and extract it in the ISO-2 or Isomerizer with the standard method using isopropyl alcohol.*

*After removing and collecting the solvent, the lettuce "opium" remains in the bottom of the reaction vessel. It may be smoked in the same manners as hash oil.*

## SUPER HASH
### *Mexican Pollen*

Dear Thai Power,

I would like to begin by thanking you for the prompt replacement of my ISO-2 which was damaged in transit. It was a drag waiting for another ten days before introducing myself to Isomerization, but the result was well worth the wait.

I believe that I have come up with a superior hashmaking method that your readers may appreciate learning.

The starting material I used was high-quality Mexican flowers from around the Lake Chapala region. The colas were about 12 inches long, well dried but still fragrant, with large, dark mature seeds. The seed hulls were bright orange, and the entire flower was so encrusted with dried resins ("pollen") that it looked as if someone had sprinkled sand on them.

As this "pollen" is actually the hashish of the plant, I decided to try to remove some of it without disturbing the beautiful and delicate flowers. I accomplished this quite simply by building a 12″ × 18″ wooden frame out of 2×2s and stretching a piece of nylon mesh cloth over this. The cloth had a consistency much like that of nylon stockings.

Out of one pound of pot I got somewhat over ½ ounce of pure fine "pollen." Even though this amount was removed from the pot, I could find no ascertainable difference in its potency before and after screening. The hash, however, was considerably stronger and tastier than the pot.

I held this over a large tray to catch the fine granules of "pollen" as they came through. I gently placed the intact flowers on top of the cloth and shook the frame. The loose pot (shake) and seeds gently rubbed on the stocking material.

I threw the "pollen" directly into the reaction vessel and added 12 ounces of alcohol. I put this into my ISO-2, put on the condensing unit, and just let it go for four hours to leach the oils into the alcohol. I then added the activator and let it run for another two hours, then neutralized and removed the solvent by collecting it in the normal manner. When the solvent was all gone I poured 12 ounces of distilled water into the reaction vessel and stirred thoroughly.

*(Editor's note: This was to dissolve and remove the baking soda neutralizer.)*

Then I dumped the water and "pollen" mix through a bedsheet and squeezed out all the water. I got a ball of hash out of that bedsheet that was so good I couldn't believe it!

## SCRAPING FOR ISO OIL

Dear TPI,

By far my favorite product from the ISO-2 is oil. For sheer power, there is no substitute for Iso oil from Columbian or Oaxacan. One good hit of this from the foil or a glass pipe, and I'm in great shape for hours.

For general all-day smoking, I have found that doublewash oil from **Reg Mex** can't be beat. Since it is so cheap to make I always smoke it in a toker pipe on ashes. No cough, great taste, good high. My big problem is how do you get the last gram out of the reaction vessel? I've tried scraping with spoons, knives, etc., and when the oil film gets thin, it just seems to move out of the way of the knife.

Sincerely,
Ron

*Dear Ron,*

*Almost all the oil can be gotten from the reaction vessel by scraping with a piece of flexible plastic about one inch wide and as thick as a credit card. A rounded corner is also helpful. This method is fast and efficient and gets all but the last traces. The last bits should be removed by heating the reaction vessel slightly and then tossing in several joints worth of grass and rubbing the grass against the side to wipe out any remaining oil.*

# ISO HASH PRESS

*Here are plans for an ISO-Hash press sent in by a reader:*

Here are some basic plans for a press that I made out of some materials that you can pick up at a metal supply house.

It is made of a square of one-inch tubing about two feet long. Then there are three pieces of hard steel about ¼ inch thick. (The reason for hard steel is so that it won't lose its shape under pressure, and it must be thick so it won't jam in the tube.)

One rod is about ¼ inch in diameter and 2½ feet long. There are two smaller rods about ⅜ inch in diameter and 2 inches long. (Make sure rods are made of a hard material so it won't bend when packing.)

Take tubing and drill two holes for the ⅜-inch rods about ½ inch from bottom and make sure the holes are straight across from each other so the base plate will sit flat on the rods. (Drill holes big enough so the rod can be removed with no effort.) Then drill a hole about ⅝ inch in diameter and about 7 inches from bottom. (That is where you load your press with ISO-Hash.)

The three pieces of ¼-inch steel are for:

1st piece:   A base that sits on top of the ⅜-inch rods.

2nd piece:   This is the plate for the plunger. Drill and tap hole for ¼-inch rod. Then thread one end of rod and cut off excess coming through plate so it is flat.

3rd piece:   A guide for the plunger rod to keep it straight while you pack. The guide is welded in the top of the tube so it won't move.

*Note:* The base and the piece for the plunger should be as close to the size of the inside of the tubing as possible so no hash will press out from around the plates, but not tight or it will get stuck.

The way it works is the plunger is raised above loading hole. Put amount of ISO-Hash to be pressed in loading hole, which falls to the base plate that is sitting on the two rods. Pack with the plunger until you feel it is packed enough. Then remove the two rods and push the whole works out of the bottom and presto — a nice cube of ISO-Hash.

## THE ALCOHOL QUESTION

Dear Thai People,

I have yet to be dissatisfied with your machine (ISO-2). I have found both instructions and operation easy to follow and perform. Because I log all operations performed with the ISO-2, I have, in my opinion and that of privileged others, been repeatedly successful in producing "certain extracts." However I am not very good in conserving the Isopropyl alcohol supplied with my machine.

To make the story short, I have been having problems finding readily available 100% Isopropyl alcohol; a money order and requisition for more is included in this letter.

The major reason for this letter is not to brag or bullshit, but to inquire if the ISO-2 can be used to distill the all-popular "rubbing alcohol," (70% isopropyl, 30% water) by your, the engineers and warrantors of the ISO-2, advice and recommendation? Please reply.

Robert
Houston, TX

*Dear Robert,*

*Rubbing alcohol can be redistilled in the ISO-2 to remove some of the water. However, you probably will not be able to facilitate an efficient isomerization. We must recommend that you use 100% Isopropyl that is available from Thai Power, Inc.*

# Appendix B

# Solvent Notes

The solvents in the table following are common in the research literature on THC. Benzene and Toluene are mentioned with respect to isomerization. Most others are used in extraction and chromatography. Petroleum ether seems to be the solvent of choice for laboratory extraction, but the consensus among amateur alchemists is that the fire hazard is too high. An examination of the flash points and N.F.P.A. health and fire hazard identification signals shows that every solvent has some serious drawback. The low-boiling solvents are almost explosive; the non-flammable solvents are more poisonous. The researcher appears to have the choice of risking death by fire or by liver and kidney damage, preceded by narcosis in both cases. One alchemist's pamphlet on cannabis oil recommends use of petroleum ether and advises storing it in the refrigerator to keep it cold. Unless the refrigerator is designed for solvent storage, this situation approximates a bomb. A refrigerator is closed to air circulation—a slight leak in the cap seal will allow the refrigerator to saturate with petroleum

(*Continued on page 92.*)

# TABLE OF SOLVENT PROPERTIES

| Name | Boiling point °C | Density g/ml |
|---|---|---|
| Benzene | 80.1 | 0.8787 |
| Chloroform | 61.2 | 1.4916 |
| Cyclohexane | 81 | 0.7791 |
| Dichloromethane | 40 | 1.335 |
| Ethanol (grain alcohol) | 78.5 | 0.7893 |
| Ethyl acetate | 77.06 | 0.9005 |
| Ethyl ether (Diethyl ether) | 34.6 | 0.714 |
| Hexane | 68 | 0.6595 |
| Isopropyl alcohol | 82.4 | 0.7851 |
| Methanol (wood alcohol) | 64.96 | 0.7914 |
| Petroleum ether | 30–60 | —— |
| Petroleum ether (commercial hexane) | 60–70 | —— |
| Toluene | 110.6 | 0.8669 |

| Water solubility | Flash Point | | N.F.P.A. Hazard Identification Signals | | |
|---|---|---|---|---|---|
| | °C | °F | health | fire | reactivity |
| δ | −17 | 2 | 2 | 3 | 0 |
| δ | — | — | non-flammable | | |
| i | −20 | −4 | 1 | 3 | 0 |
| δ | — | — | 2 | 0 | 0 |
| ∞ | 13 | 55 | 0 | 3 | 0 |
| s | 4 | 24 | 1 | 3 | 0 |
| s | −45 | −49 | 2 | 4 | 1 |
| i | −30 | −22 | 1 | 3 | 0 |
| ∞ | 12 | 53 | 1 | 3 | 0 |
| ∞ | 11 | 52 | 1 | 3 | 0 |
| i | −57 | −70 | 1 | 4 | 0 |
| i | −32 | −25 | 1 | 4 | 0 |
| i | 4 | 40 | 2 | 3 | 0 |

ether vapor (flash point −32°C/−24°F); a spark from the motor or defroster could then ignite the confined mixture.

The effect of the type of solvent on the stability of marijuana solutions to decomposition has also been recently reported (Bonuccelli, 1979). Chloroform (and probably other chlorine-containing solvents) promotes significant decomposition of THC on storage; this decomposition is speeded up by light. In sunlight, a chloroform solution lost 25–35% of its THC in 30 minutes. The author found ethanol better for storage, but slight decomposition still occurred. Storage in cold and darkness is recommended. Keeping samples under inert atmosphere (nitrogen) will also retard decomposition.

**FLASH POINT** The temperature at which the material will give off enough vapor in air to form an ignitible mixture that will propagate flame. These are typical test values by closed cup test, except when designated "OC" open-cup.

**WATER SOLUBILITY** Symbols used to indicate the relative water solubility of the substances:
∞  miscible
v   more than 50 grams dissolve in 100 milliliters of water
s   5 to 50 grams dissolve in 100 milliliters of water
δ   less than 5 grams dissolve in 100 milliliters of water
i   insoluble
d   decomposes on contact with water

**NFPA HAZARD IDENTIFICATION SIGNALS (From Identification System for Fire Hazards of Materials (NFPA No. 704-M))**
IDENTIFICATION OF HEALTH HAZARD — Type of Possible Injury: (Color Code: BLUE)
4 Materials which on very short exposure could cause death or major residual injury even though prompt medical treatment were given.
3 Materials which on short exposure could cause serious temporary or residual injury even though prompt medical treatment were given.
2 Materials which on intense or continued exposure could cause temporary incapacitation or possible residual injury unless prompt medical treatment is given.
1 Materials which on exposure would cause irritation but only minor residual injury even if no treatment is given.
0 Materials which on exposure under fire conditions would offer no hazard beyond that of ordinary combustible materials.

IDENTIFICATION OF FLAMMABILITY — Susceptibility of Materials to Burning: (Color Code: RED)
4 Materials which will rapidly or completely vaporize at atmospheric pressure and normal ambient temperature, or which are readily dispersed in air and which will burn readily.

3 Liquids and solids that can be ignited under almost all ambient temperature conditions.

2 Materials that must be moderately heated or exposed to relatively high ambient temperatures before ignition can occur.

1 Materials that must be preheated before ignition can occur.

0 Materials that will not burn.

IDENTIFICATION OF REACTIVITY (STABILITY)—Susceptibility to Release of Energy: (Color Code: YELLOW)

4 Materials which are readily capable of detonation or of explosive decomposition or reaction at normal temperatures and pressures.

3 Materials which are capable of detonation or explosive reaction but require a strong initiating source or which must be heated under confinement before initiation or which react explosively with water.

2 Materials which are normally unstable and readily undergo violent chemical change but do not detonate. Also materials which may react violently with water or which may form potentially explosive mixtures with water.

1 Materials which are normally stable, but which can become unstable at elevated temperatures and pressures or which may react with water with some release of energy but not violently.

0 Materials which are normally stable, even under fire exposure conditions, and which are not reactive with water.

Adapted from
*Handbook of Laboratory Safety*
2nd ed., Chemical Rubber Co.
(Cleveland, Ohio, 1971)

# Glossary

**Acetic anhydride** $(CH_3CO)_2O$ — Compound produced from two molecules of acetic acid by removal of one water molecule. Used for acetylation.

**Acetylation** — The introduction into an organic compound of an acetyl group $(CH_3CO)$. Acetic anhydride can be used to do this; it reacts with -OH and -NH$_2$ groups on the molecule.

**Alchemy** — An ancient practice of chemistry, some goals of which were the transformation of base metals into gold and discovery of the elixir of life. The writings of the alchemists have also been given philosophic and religious interpretations, an example being C. G. Jung's interpretation of dreams as alchemical symbolism.

**Beam test** — A chemical test which gives a violet color if cannabidiol is present. The test gives no such color with THC or cannabinol.

**Cannabidiol** — A non-psychoactive cannabinoid (compound related to THC) in marijuana which by elimination of water and formation of a third ring structure can be converted to THC.

**Cannabidiol-bis-3,5-dinitrobenzoate** — The derivative of cannabidiol formed by reaction with 3,5-dinitrobenzoyl chloride. It crystallized more readily than cannabidiol, and this is used to effect a purification. After crystallization the derivative is converted back to cannabidiol.

**Cannabinol** — A non-psychoactive cannabinoid (compound related to THC) which is a product of the oxidation of THC.

**Cannabis** — Genus name of the marijuana plant; sometimes refers to the plant in general or its products.

*Cannabis indica* — Oriental strain (perhaps a separate species, though it produces fertile seeds with *Cannabis sativa*) of marijuana. Usually shorter with broader leaf blades and higher THC content than *Cannabis sativa*.

**Cannabis oil** — The viscous liquid resulting from the extraction of cannabis with an organic solvent and the subsequent removal of that solvent.

*Cannabis sativa* — Strain of marijuana originally cultivated for fiber rather than resin. Usually has a higher cannabidiol to THC ratio than *indica*, although some varieties are quite potent.

*Charas* — Hindu word for hashish, the pressed resin of marijuana, as distinguished from manicured tops *(ganja)* or loose marijuana debris *(bhang)*.

**Chloroform** ($CHCl_3$) — A non-flammable organic solvent. Often used in the past as a general anesthetic, it has now been replaced by less toxic chemicals.

**Chromatography** — An analytical technique for the separation of different compounds from each other. A number of methods (paper, thin layer, solid-liquid, and gas chromatography) operate on the same basic principle. A mobile phase moves through a stationary phase, and when a mixture of compounds is introduced into the flow, different compounds move at differing rates depending on their affinity for one phase or the other.

**Claisen flask** — A glass flask with a U-shaped neck, used for distillation.

**Condenser** — An apparatus, cooled by circulating cold water, used to condense vapors of boiling liquids back to liquids.

**Countercurrent extraction** — A liquid-liquid extraction process in which two immiscible liquids move past each other in opposite directions. Similar in principle to chromatography.

**Crystallization** — The formation of a solid in crystal form from a saturated liquid solution. Crystals include mainly the molecules of the pure compound, so when the liquid is removed from the crystals a purification usually results.

**3,5-dinitrobenzoyl chloride** (($NO_2$)$_2C_6H_3COCl$) — The chemical used to form 3,5-dinitrobenzoate derivatives of organic molecules. These derivatives usually crystallize more easily than the original compounds and are used to purify or identify them.

**Distillation** — The conversion of a liquid to a vapor and later condensation of the vapor to liquid in a different container. Often used to purify solvents.

**Emulsion layer** — A cloudy region at the boundary of two immiscible liquids where the two liquids are present together, one as fine bubbles in the other.

**Erlenmeyer flask** — A conical flask with a flat bottom and a short, narrow, straight-sided neck. Named after Emil Erlenmeyer, a German chemist.

**Essential oils** — The volatile oils present in fruits, flowers, leaves, etc., which give them characteristic odors.

**Ethanol** ($CH_3CH_2OH$) — Organic solvent (also known as ethyl alcohol or grain alcohol) of low toxicity, usually denatured (made unfit to drink) by the addition of small amounts of toxic chemicals.

**Extraction** — A method which removes a desired material from a mixture of others. Here, the use of a solvent which dissolves THC and other compounds but not the insoluble compounds (cellulose, protein, etc.).

**Flashing off** — The removal of a solvent by reducing the pressure in the container and causing the liquid to vaporize. The vapor can then be condensed at a cooler temperature in another vessel.

**Florisil** — A material used as a stationary (nonmoving) phase of liquid-solid chromatography.

**Fractional distillation** — A distillation process which separates liquids of differing boiling points.

**Hashish** — Pressed marijuana resin. The word is derived from the Arabic name for members of Hasan ibn as-Sabbāh's cult, which is also the root of the word *assassin.*

**Isomer** — In chemistry, a compound with the same atomic composition as another compound but with differing properties.

**Isomerization** — The formation of an isomer. Here, the formation of THC from cannabidiol.

**Isopropyl alcohol** ($(CH_3)_2CHOH$) — Organic solvent that is the main constituent of common rubbing alcohol.

**Lettuce opium** — Popular, openly sold extract of lettuce; reportedly psychoactive, though this has not been scientifically substantiated.

**Methanol** ($CH_3OH$) — A toxic organic solvent also known as methyl alcohol or wood alcohol. It is sometimes used to denature ethyl alcohol.

**Neutralizing** — The addition of a base to an acid solution (or vice versa) to the point that the solution is neither acidic nor basic (pH = 7.0). Sometimes when "neutralizing an acid" is referred to, the final solution is not really neutral but basic.

**Oiler** — see reefer.

**Parr bomb** — An apparatus designed to determine the caloric value (the amount of heat produced by burning) of fuels. Here used as a vessel for carrying out a reaction at high pressure. Named after Samuel Wilson Parr, an American chemist.

**Petroleum ether** — A highly flammable fraction of petroleum, distilling between 40° and 70°F. It is composed mainly of pentanes and hexanes (molecules with chains of carbon atoms five and six atoms long). It is frequently used in cannabis research as a solvent for THC, but its high fire hazard makes it dangerous to use.

**Psychoactive and non-psychoactive** — Indicating the presence or absence of specific mental effects. In cannabis, psychoactive compounds (mainly THC) are responsible for the characteristic euphoric state.

**Purification** — Generally, a method which increases the percentage of a desired substance. Here, the use of two immiscible liquids to remove impurities from THC.

**Reefer** — Early 20th-century slang for marijuana cigarette. Repopularized in the '60s. Sometimes used in the '70s to describe a marijuana cigarette which has been impregnated with cannabis extract; also called an "oiler."

**Refluxing** — The continuous return of condensed vapors to the boiling container from which they came.

**Resin** — The oily secretion of plants. In marijuana, the THC is located mostly in the resin; however, high resin content is not synonymous with high THC content.

**Rheostat** — A mechanism for adjusting the resistance of an electric circuit. Often used in labs to adjust the intensity of heaters or lights, and to regulate the rates of electric motors.

**Rotating (higher, lower; also specific rotation)** — The number of angular degrees through which a plane of polarized light is rotated by passing through a solution with a concentration of 1 gram per cubic centimeter in a tube 1 decimeter ($1/10$ meter) long. Higher and lower rotating forms are comparative terms for two isomers with higher and lower values of specific rotation.

**Safety box** — A container which shields the worker from contact and possible harm from explosion during certain chemical procedures. The atmosphere in the box can be replaced with inert gas to prevent combustion.

**Solvent** — In general, a liquid which is effective in dissolving a material; in particular, an organic compound which is liquid and dissolves THC readily.

**Soxhleting** — A process of continuous solvent extraction of a solid. The Soxhlet apparatus is a combination of a distillation flask, a reflux condenser, and a vessel (which holds the solid) connected to a siphon. Named after the German chemist, Franz von Soxhlet.

**THC** — Tetrahydrocannabinol, the main psychoactive compound in marijuana.

**THC acetate** — The compound produced by acetylation of THC. Its psychoactive properties differ from THC, but the nature of the difference is still controversial.

# References

Adams, R. et al. "Structure of Cannabidiol, a Product Isolated from the Marihuana Extract of Minnesota Wild Hemp, I." *Journal of the American Chemical Society* 62 (1940): 196–200.

Adams, R. et al. "Structure of Cannabidiol. XII: Isomerization to Tetrahydrocannabinols." *Journal of the American Chemical Society* 63 (1941): 2209–13.

Bonuccelli, C. M. "Stable Solutions for Marijuana Analysis." *Journal of Pharmaceutical Sciences* 68 (1979): 262–63.

Davis, K. H. et al. "The Preparation and Analysis of Enriched and Pure Cannabinoids from Marihuana and Hashish." *Lloydia Journal of Natural Products* 33 (1970): 453–60.

Drug Enforcement Administration, *Drug Enforcement*, 1973.

Gaoni, Y., and Mechoulam, R. "Isolation, Structure, and Partial Synthesis of an Active Constituent of Hashish." *Journal of the American Chemical Society* 86 (1964): 1646–47.

Hively, R. L. et al. "Isolation of *trans*-$\Delta^6$-Tetrahydrocannabinol from Marijuana." *Journal of the American Chemical Society* 88 (1966): 1832–33.

Mechoulam, R. "Marihuana Chemistry." *Science* 168 (1970): 1159–66.

Mechoulam, R. *Marijuana*. New York: Academic Press, 1973.

Starks, M. *Marijuana Potency*. Berkeley: And/Or Press, 1977.

Steere, N. V., ed. *Handbook of Laboratory Safety*. 2nd ed. Cleveland: Chemical Rubber Co., 1971.

# Index

Tetrahydrocannabinol. *See* THC

Texas Super Hash, 67–69, 80

THC, 17–23, 55; appearance, 59; in chromatography, 58, 59; conversion from low- to high-rotating forms, 20–23, 73; decomposition, 3, 19–20, 72, 92; in essential oils, 5, 17; isomerization, 17–23, 28, 55; rotation range, 61. *See also* Acetylation; THC acetate; THC, pure; etc.

THC, pure, 55; conversion from pure cannabidiol, 61

THC solutions, decomposition, 19–20, 92; storage of, 19, 92

THC acetate, potency, 24; weight increase, 24

THC acetate, synthesis, 24–36; distillation, 32, 33–36; equipment, 24–36; flow chart, 32; purification, 36; refluxing, 28–33; safety, 24–28

Toluene, 19, 60, 89, 90–91

Translucent oil, 43, 55

Vaporization. *See* Smoking and smoking methods

Water bath. *See* Evaporation; Refluxing

Water solubility, definition, 93

Waxes, 74, 75. *See also* Resins

Wood alcohol. *See* Methanol